**Fire Safety Aspects
of
Polymeric Materials**

VOLUME 9
SHIPS

Report of

The Committee on Fire Safety
Aspects of Polymeric Materials

NATIONAL MATERIALS ADVISORY BOARD
Commission on Sociotechnical Systems
National Research Council

Publication NMAB 318-9
National Academy of Sciences
Washington, D.C.
1980

VOLUME 9 — SHIPS
Fire Safety Aspects of Polymeric Materials

Printed in U.S.A.
Library or Congress Card No. 77-79218
ISBN 087762-230-2

NOTICE

The project that is the subject of this report was approved by the Governing Board of the National Research Council, whose members are drawn from the councils of the National Academy of Sciences, the National Academy of Engineering, and the Institute of Medicine. The members of the Committee responsible for the report were chosen for their special competences and with regard for appropriate balance.

This report has been reviewed by a group other than the authors according to procedures approved by a Report Review Committee consisting of members of the National Academy of Sciences, the National Academy of Engineering, and the Institute of Medicine.

This study by the National Materials Advisory Board was conducted under Contract No. 4-35856 with the National Bureau of Standards.

Printed in the United States of America.

FOREWORD

This volume is one of a series of reports on the fire safety aspects of polymeric materials. The work reported here represents the results of the first in-depth study of this important subject. The investigation was carried out by a committee of distinguished polymer and fire technology scholars appointed by the National Academy of Sciences and operating under the aegis of the National Materials Advisory Board, a unit of the Commission on Sociotechnical Systems of the National Research Council.

Polymers are a large class of materials, most new members of which are man-made. While their versatility is demonstrated daily by their rapidly burgeoning use, there is still much that is not known or not widely understood about their properties. In particular, the burning characteristics of polymers are only now being fully appreciated and the present study is a landmark in the understanding of the fire safety of these ubiquitous materials.

In the first volumes of this series the committee has identified the limits of man's knowledge of the combustibility of the growing number of polymeric materials used commercially, the nature of the by-products of that combustion, and how fire behavior in these systems may be measured and predicted. The later volumes deal with the specific applications of polymeric materials, and in all cases the committee has put forth useful recommendations as to the direction of future actions to make the use of these materials safer for society.

Harvey Brooks, Chairman
Commission on Sociotechnical Systems

ABSTRACT

This is the ninth volume in a series that examines fire safety aspects of polymeric materials, with primary emphasis on human survival. Other volumes in the series deal with materials: state of the art; test methods, specifications, and standards; special problems of smoke and toxicity; fire dynamics and scenarios; aircraft: civil and military; and applications to building, land transportation vehicles; and mines and bunkers. An executive summary volume (Elements of Polymer Fire Safety and Guide to the Designer) has been added to the series.

In this volume, the fire safety aspects of polymeric materials used in ships, boats, craft, and devices — both for commercial and military service — are examined, with primary focus on improving fire safety. The study addresses (a) the physical and chemical parameters that influence flammability, smoke, and toxicity; (b) the physical and chemical material combinations that are used; (c) the use of these materials in devices, subsystems and systems; (d) the geometry, position, and environments of the material, and (e) the contribution of the materials to system performance in normal and abnormal modes under fire conditions.

Included in each chapter are conclusions and recommendations based on primary considerations of human survival, and secondary consideration of system survival. The major conclusions and recommendations are extracted and combined in Chapter 2, but the reader is advised to consult the conclusions and recommendations in Chapters 3 through 10 for the committee's complete views.

VOLUMES OF THIS SERIES

Volume 1   Materials: State of the Art

Volume 2   Test Methods, Specifications, Standards

Volume 3   Smoke and Toxicity

Volume 4   Fire Dynamics and Scenarios

Volume 5   Elements of Polymer Fire Safety and Guide to the Designer

Volume 6   Aircraft: Civil and Miltary

Volume 7   Buildings

Volume 8   Land Transportation Vehicles

Volume 9   Ships

Volume 10   Mines and Bunkers

PREFACE

The National Materials Advisory Board (NMAB) of the Commission on Socio-technical Systems, National Research Council, was asked by the Department of Defense Office of Research and Engineering and the National Aeronautics and Space Administration to "initiate a broad survey of fire-suppressant polymeric materials for use in aeronautical and space vehicles, to identify needs and opportunities, assess the state of the art in fire retardant polymers (including available materials, pioducts, costs, data requirements, methods of test, and toxicity problems), and describe a comprehensive program of research and development needed to update the technology and accelerate application where advantages will accrue in performance and economy."

In accordance with its usual practice, the NMAB convened representatives of the requesting agencies and other agencies working in the field to determine how the project might best be undertaken. It was quickly apparent that widepsread duplication of interest exists. At the request of other agencies, sponsorship was made available to all government departments and agencies having an interest in fire safety. Concurrently, the scope of the project was broadened to take into account the needs of the new sponsors, as well as those of the original sponsors.

In addition to the Department of Defense and the National Aeronautics and Space Administration, sponsors of this study now include the Departments of Agriculture, Commerce (National Bureau of Standards), Interior (Bureau of Mines, Division of Mining Research, Health, and Safety), Housing and Urban Development, Health, Education and Welfare (National Institute of Occupational Safety and Health), Transportation (Federal Aviation Administration, Coast Guard), and Energy, as well as the Consumer Product Safety Commission, Environmental Protection Agency, Postal Service, and the National Fire Prevention and Control Administration.

The committee was originally constituted on November 30, 1972. The membership was expanded to its present status on July 25, 1973, after presentation of reports by liaison representatives which covered needs, views of problem areas, current activities, future plans, and relevant resource materials. Tutorial presentations were made at meetings at the Academy and during site visits, when the committee or its panel met with experts and organizations concerned with fire safety aspects of polymeric materials. These site visits (upwards of a dozen) were an important feature of the committee's search for authentic information. Additional inputs of foreign fire technology were supplied by the U.S. Army Foreign Science and Technology Center and NMAB staff.

ACKNOWLEDGEMENTS

Panel members of the National Materials Advisory Board Committee on Fire Safety Aspects of Polymeric Materials, as well as government liaison representatives, drafted this report, which the entire committee reviewed and finalized. Conclusions and recommendations are the sole responsibility of the committee.

Coordination of this volume was performed by Dr. R. R. Hindersinn. Panel members responsible for the first draft of various chapters include Rear Admiral W. C. Hushing, U.S.N. (Retired), Dr. R. S. Magee, Dr. A. R. Gilbert, Mr. I. Litant, Lt. Col. E. B. Altekruse, M.C., U.S.A., Mr. M. Soldo, Mr. D. Smillie, Dr. A. Rosenthal, Mr. L. D. Polland and Dr. R. R. Hindersinn.

Liaison representatives who also made substantial and definitive contributions in the preparation of various part of this volume include Dr. George Thomas, Department of the Army, Army Materials and Mechanics Research Center, Watertown, Mass.; Mr. Daniel Pratt, Naval Ships Engineering Center, Prince George's Center, Hyattsville, Md.; Mr. Jack Ross, Wright-Patterson Air Force Base, Ohio; and Mr. Charles Bogner, Naval Ship Engineering Center, Department of the Navy, Washington, D.C.

Mr. Harvey Paige, Executive Secretary, Maritime Administration Research Board, National Academy of Sciences, also provided valuable contributions to this volume.

Excellent technical presentations were made by a number of government organizations; those made by the Navy and Coast Guard were most directly helpful.

The Bath Iron Works, a division of the Congoleum Corporation, provided outstanding demonstrations of commercial and naval ships under construction as well as helpful technical discussions and comments.

Particular thanks are due to Dr. George Thomas and Ms. Betty Sterling of the Army Materials and Mechanics Research Center (AMMRC) for their administrative support and other assistance during many panel meetings held at AMMRS.

I acknowlege with gratitude the assistance in this project of Dr. Robert S. Shane, NMAB Consultant.

Dr. Herman F. Mark, Chairman

NATIONAL MATERIALS ADVISORY BOARD

Chairman

Mr. William D. Manly
Senior Vice President
Cabot Corporation
Boston, Massachusetts

Past Chairman

Mr. Julius J. Harwood
Director, Materials Science Laboratory
Engineering and Research Staff
Ford Motor Company
Dearborn, Michigan

Members

Dr. George S. Ansell
Dean, School of Engineering
Rensselaer Polytechnic Institute
Troy, New York

Dr. H. Kent Bowen
Professor, Ceramic and Electrical
  Engineering
Massachusetts Institute of Technology
Cambridge, Massachusetts

Dr. Van L. Canady
Senior Planning Associate
Mobil Chemical Company
New York, New York

Dr. George E. Dieter, Jr.
Dean, College of Engineering
University of Maryland
College Park, Maryland

Dr. Joseph N. Epel
Director, Plastics Research and
  Development Center
Budd Corporation,
Troy, Michigan

Dr. Larry L. Hench
Professor and Head, Ceramics Division
Department of Materials Science and
  Engineering
University of Florida
Gainesville, Florida

Dr. Robert E. Hughes
Professor of Chemistry,
Executive Director
Materials Science Center, Department
  of Chemistry
Cornell University
Ithaca, New York

Dr. John R. Hutchins III
Vice President and Director of
  Research and Development
Technical Staff Division
Corning Glass Works
Corning, New York

Dr. Sheldon E. Isakoff
Director, Engineering Research and
  Development Division
E. I. DuPont de Nemours & Co., Inc.
Wilmington, Delaware

Dr. Frank E. Jaumot, Jr.
Director of Advanced Engineering
Delco Electronics Division
General Motors Corporation
Kokomo, Indiana

Dr. James W. Mar
Professor, Aeronautics and
  Astronautics
Massachusetts Institute of Technology
Cambridge, Massachusetts

GIULIANA C. TESORO, Adjunct Professor, Massachusetts Institute of Technology, 278 Clinton Avenue, Dobbs Ferry, New York.

NICHOLAS W. TSCHOEGL, Professor Department of Chemical Engineering, Division of Chemistry and Chemical Engineering, California Institute of Technology, Pasadena California.

ROBERT B. WILLIAMSON, Associate Professor of Engineering Science, Dept. of Civil Engineering, University of California, Berkeley, California.

Liaison Representatives

JEROME PERSH, Staff Specialist for Materials and Structures, ODUSDRE(ET), Department of Defense, Washington, D.C.

ERNEST B. ALTERKRUSE, Chief, Department of Clinics, Moncrief Army Hospital, Fort Jackson, Columbia, South Carolina.

ALLAN J. McQUADE, U.S. Army Natick Laboratories, Natick, Massachusetts.

GEORGE R. THOMAS, Department of the Army, Army Materials and Mechanics Research Center, Watertown, Massachusetts.

HERBERT C. LAMB, Naval Facilities Engineering Command, Alexandria, Virginia.

DANIEL PRATT, Naval Ships Engineering Center, Prince George's Center, Hyattsville, Maryland.

GEORGE SORKIN, Naval Sea Systems Command, Washington, D.C.

JACK ROSS, Wright-Patterson Air Force Base, Ohio

BERNARD ACHHAMMER, National Aeronautics and Space Administration, NASA Headquarters, Advanced Research and Technology Division, Washington, D.C.

JOHN A. PARKER, National Aeronautics and Space Administration, Ames Research Center, Moffett Field, California

ARNOLD WEINTRAUB, Energy Research and Development Administration, Washington, D.C.

DAVID FORSHEY, Bureau of Mines, Washington, D.C.

JOSEPH E. CLARK, National Fire Prevention Control Agency, Dept. of Commerce, Washington, D.C.

CLAYTON HUGGETT, Center for Fire Research, National Bureau of Standards, Washington, D.C.

LESLIE H. BREDEN, Fire Safety Engineering Division, Center for Fire Research, National Bureau of Standards, Washington, D.C.

IRVING LITANT, Department of Transportation (DoT), Systems Center, Kendall Square, Cambridge, Massachusetts.

GEORGE BATES, JR., Department of Transportation, Federal Aviation Administration, Washington, D. C.

CONTENTS

Appendices

CHAPTER 1

INTRODUCTION

1.1 Scope and Methodology of the Study

The charge to the NMAB Committee on Fire Safet Aspects of Polymeric Materials was set forth in presentations made by the sponsoring agencies. Early in its deliberations, however, the committee concluded that its original charge required modification and expansion if the crucial issues were to be fully examined and the needs of the sponsoring organizations met. Accordingly, the committee agreed it would direct its attention to the behavior of polymeric materials in a fire, with special emphasis on human safety considerations. Excluded from consideration were therapy after fire-caused injury and mechanical aspects of design not related to fire safety. The work of the committee includes (1) a survey of the state of pertinent knowledge, (2) identification of gaps in that knowledge, (3) identification of work in progress, (4) evaluation of the work as it relates to the identified gaps, (5) development of conclusions, (6) recommendations for action by appropriate public and private agencies, and (7) estimation, when appropriate, of the benefits that might come from implementation of the recommendations. Within this framework, functional areas were addressed as they relate to specific situations; end uses were considered when fire was a design consideration and the end uses were of concern to the sponsors of this study.

Attention was given to natural and synthetic polymeric materials, primarily in terms of their composition, structure, relation to processing, and geometry (i.e., film, foam, fiber, etc.), but their incorporation into an end-use component or structure also was considered. Test methods, specifications, definitions, and standards were considered. Regulations, hwoever, were dealt with only in relation to end uses.

The products of combustion, including smoke and toxic substances, were considered in terms of their effects on human safety; morbidity and mortality were treated only as a function of the materials found among the products of combustion. The committee considered potential exposure to fire-retardant polymers (including skin contact) in situations not including pyrolysis and combustion. This was done in relation to various end uses.

In an effort to clarify the understanding of the phenomena accompanying fire, consideration was given to the mechanics of mass and energy transfer (fire dynamics). The opportunity to develop one or more scenarios to guide thinking was provided; however, as noted above, firefighting was given only superficial consideration. To assist those who might use natural or synthetic polymers in components or structures, consideration also was given to design principles and criteria.

In organizing its work, the committee concluded that its analysis of the fire safety of polymeric materials should address the materials themselves, fire dynamics, and the large societal systems affected. This decision led to separate treatment of the technical-functional aspects of the problem and the aspects of product end use.

Accordingly, the committee is presenting its findings in five disciplinary and five end-use reports:

Volume 1 Materials: State of the Art
Volume 2 Test Methods, Specifications, and Standards
Volume 3 Smoke and Toxicity
(Combustion Toxicology of Polymers)
Volume 4 Fire Dynamics and Fire Scenarios
Volume 5 Elements of Polymer Fire Safety and
Guide to the Designer
Volume 6 Aircraft (Civil and Military)
Volume 7 Buildings
Volume 8 Land Transportation Vehicles
Volume 9 Ships
Volume 10 Mines and Bunkers

1.2 Scope and Limitations of This Report

This volume specifically examines the polymeric materials used in ships, boats, craft, and devices for both commercial and military service. For each category of vessel the committee has attempted to determine:

1. The parameters, physical and chemical, that influence flammability, smoke, and toxicity.
2. The material combinations, physical and chemical, that are used.
3. The use of the materials in devices, subsystems, and systems.
4. The geometry, position, and environment of the material.
5. The contribution of the materials to system performance in normal and abnormal modes (fire).

Since much knowledge needed to make such determinations was lacking, the judgments of the committee are tentative and subject to revision; they represent only "best possible estimates" based on what is currently known. (Information on vessels was delivered to the committee before September 1977 and literature references beyond that date are not included).

Although the relative priority of conclusions and recommendations was part of the committee's discussions, this report does not attempt to advise managers of resources how to allocate them, vis-a-vis other demands.

Specific polymeric materials generally excluded from this report include propulsion fuels and hydraulic fluids. In addition, in commercial ships where the cargo is so widely varying in content and composition and in military ships where the composition and amount of payload (offensive power) is classified, no attempt has been made to consider them. It should be noted that unclassified information on relevant materials is contained in Volume 1 — Materials: State of the Art. The fire load associated with passenger and crew carry-on material has not been addressed although the significance of this material is not minimized.

The committee assessed polymeric materials used in vessels relative to:

1. Current materials knowledge and data.
2. Current test methods and standards.
3. Real world fire environment.
4. State of knowledge of smoke and toxicity.
5. Systems applications.
6. Potential for improvements.

1.3 Committee Viewpoints

Members of the committee are knowledgeable about materials, research and development, applications, system design, and evaluation; liaison representatives deal with research and development, regulation, procurement, operations, and analysis. Thus, each material and its problems were evaluated by experts.

It is necessary to emphasize that each of the many statements concerning fire safety aspects of polymeric materials applies only to specific situations. Statements in this volume must not be taken out of context and applied to the use of identical materials in other situations. The reader is further cautioned that many polymeric materials are poorly identified; the acceptance of a manufacturer's nomenclature is often not sufficient to determine what, in fact, the material is.

1.4 Methodology

By the time this volume was embarked upon the committee's expertise and experience had been enhanced by three years of close association and communication and by the preparation of the five disciplinary reports and two of the end-use reports. This experience was very helpful in addressing the fire safety aspects of polymeric materials used in ships and water craft.

In addition to assembly and analysis of data and information (including regulation and specifications), the committee sought and obtained tutorial presentations from the principal government agencies involved, the Coast Guard (for commercial vessels) and the Navy. Committee members visited a private shipyard to observe commercial and Navy vessels under construction and discuss with shipbuilders their problems with polymeric materials.

1.5 General Considerations

Safety of life and limb of those in water-borne vehicles was the first priority in the committee's considerations; discussions, conclusions, and recommendations are heavily weighted in that direction. While there is no doubt that the rapidly escalating costs of vessels dictate that substantial effort be directed towards reduction of property damage from fire, this report assigns a secondary priority to such matters.

The general public has little exposure to fire danger on U.S. commercial ships, since there are very few which carry passengers; only a small crew is generally involved on such ships. U.S. Navy ships carry no public passengers, but generally have substantial crews. Although the number of small pleasure craft is very large, the fatalities and injuries from fires is surprisingly low.

Despite the relatively small impact to date of polymers upon ship construction, the economic pressures for the use of these newer materials in vessels remain great because of their inherent advantages. Some of the more important of these advantages are reductions in cost, weight, and maintenance. These advantages result from the superior resistance to corrosion and soiling, significantly reduced fabrication costs, and reduced energy costs both in materials production and systems operations. As indicated in Figure 1.1, the energy requirements for the production of a unit of polymer are only one tenth to one half that for such metals as aluminum or magnesium, while energy savings of from 10 to 80 percent are possible relative to steel, depending upon the polymer chosen. This advantage of polymers over metals also extends into energy consumption, particularly in high-speed, special-purpose vessels, because the lower density of polymers reduces the weight of the finished product with a resulting saving in energy needed to propel the ship through the water.

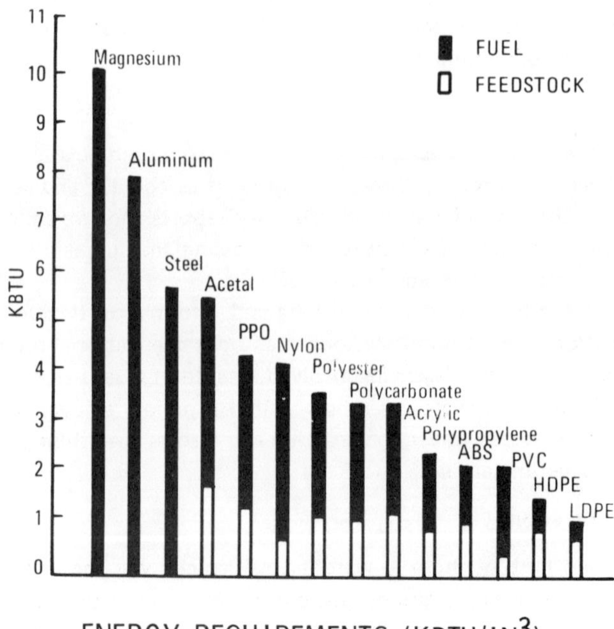

ENERGY REQUIREMENTS (KBTU/IN.^{3})

Figure 1.1 Energy requirements (kbtu/in.^{3})

Despite the advantages of expanded use of polymers in ship construction, the ready flammability of most of the low-cost, high volume materials now produced makes the close control of these materials imperative.

The Coast Guard, in setting specifications for commercial ships, craft, boats, and devices, and the Navy, in setting specifications for its own vessels, have used a system approach and the latest information on polymer performance. As a result, a high degree of fire safety has been attained.

REFERENCES

P. F. Brake, "Potential Tank-Automotive Applications for Organic Materials, Part II — Composite Materials," Army Materials and Mechanics Research Center Report, Mar. 31, 1976. U.S. Army Materials and Mechanics Research Center, Watertown, MA. 02172.

CHAPTER 2

SUMMARY, CONCLUSIONS AND RECOMMENDATIONS

2.1 Summary

This report discusses the materials, tests, regulations, specifications, and special considerations as applied to fire safety aspects of polymeric materials used in such applications as hulls, fittings, and furnishings used to construct and control various kinds of watercarft. Naval and commercial craft are included, as are special devices such as oil-drilling rigs, ocean mining vessels, and high-performance ships, e.g., surface effect and hydro-foil craft. The latter have close ties to aircraft manufacturing and operating practices.

The fire safety record of all watercraft is very good and is improving as new international safety-at-sea regulations are implemented. Fire statistics indicate that the fire hazard on ships is not acute when compared to fire-related injuries, deaths, and property damage in land transportation vehicles or buildings. A few specific problems deserve additional emphasis and higher priority, (e.g., electrical wiring insulation). In general, early and orderly application of new knowledge of polymers by the Coast Guard and the Navy will continue to support a superior fire safety record (compared with land transport and building usage of synthetic polymers).

2.2 General Conclusions and Recommendations

Conclusion: Manmade polymers offer substantial advantages when they are properly substituted for metals in structures and components. These advantages include reduction of weight, reduction of cost to manufacture and form, reduction of energy required to produce and manufacture, and improved corrosion resistance. However, in some cases, the effect on fire safety arising from this substitution has been undesirable. *Recommendation:* The Navy and Coast Guard should establish aggressive programs to classify and/or develop polymer compositions and composites from a fire safety point of view, suitable for substitution for metals in ship structures and equipments. This recommendation will require substantial additional investment in manpower, funds and laboratory capability.

2.3 Conclusions and Recommendations. Fire Dynamics and Scenarios (Chapter 3)

Conclusions based on the scenario of the SS C.V. Sea Witch-SS Esso Brussels collision and fire (See Chapter 3, Section 3.4.2):

1. The ability of the crew of the SS C.V. Sea Witch to survive in the after deckhouse for a period of about 1 hour, and the electrician for about 2 hours in the emergency generator room, while the stern of the vessel was engulfed in flames, can be attributed to the excellent structural fire resistance of the interior paneling and furnishing of the vessel.

2. The S.S. Esso Brussels deckhouse materials of construction and interior furnishings were essentially combustible materials; they were almost completely con-

sumed by the fire. The complete and rapid spread of the fire through the living spaces of both deckhouses emphasizes the absence of structural fire protection aboard this vessel. Had the crew sought shelter aboard, they probably would have died in the deckhouse fire. The hazards of combustible construction within the living accommodations of a tank vessel were clearly demonstrated in this casualty.

3. The rapid spread of smoke through the engine room and after-quarters complicated the survival efforts of the crew of the SS C.V. Sea Witch. The early abandonment of the engine room because of the dense smoke could have been a significant factor under different circumstances in which positive steps to combat the fire or maneuver the vessel were undertaken by the crew.

4. The deck cargo containers on the side of the vessel were first affected by the flames and heat surrounding the SS C.V. Sea Witch. The fire accelerated as containers ruptured or were consumed during the fire, exposing additional fuel to the fire. The progress of the deck fire was uninhibited either by the small separation space between bays or the boundaries provided by the containers. The absence of any effective fire stops to separate the deck cargo mass — some 10 feet high, more than 320 feet in length, and extending the width of the vessel — compounded the problem of fire containment.

5. The hulls of both vessels remained intact after being subjected to intense heat; this substantiates the superiority of steel as a structural material, especially when compared to the extensive damage to the nonferrous fittings on the SS Esso Brussels and aluminum containers on deck of the SS C.V. Sea Witch.

Recommendations from Analysis of the Scenario:

1. Check the adequacy of existing container construction standards and the possible need for additional shipboard fire protection standards in view of the rapid spread of fire through the containers.

2. Conduct further studies to develop methods to prevent the spread of smoke within the interior of burning vessels. Ventilation systems should be designed to provide manual or automatic means, not only to prevent the spread of fire, but also the spread of smoke.

Conclusion: Statistics indicate that the fire problem on ships is not acute when compared to fire-related injuries, deaths, and property damage in land transportation vehicles or buildings. Although each ship fire is different, the development and analysis of fire scenarios leads to the identification of common elements. This allows identification of the critical states in fire development, suggests opportunities for fire prevention, and directs attention towards various methods for control. *Recommendation:* Develop a wide spectrum of generalized ship fire scenarios, as outlined in Chapter 3, based on real or credible incidents. The specific fire dynamic elements in these scenarios (e.g., rate of fire spread, rate of heat release, etc.) should be further quantified when necessary by information obtained from full-scale experiments. The ultimate goal is to develop the capability of employing small-scale experiments that can be used to predict full-scale events.

Conclusion: Scenarios should be prepared so as to permit generalization from the particular incident described. A good scenario provides a basis for exploration of alternative paths of fire initiation and growth and for analysis of the effect on fire safety performance of changes in specifications, materials, and design. *Recommendation:* Use scenarios for the analysis of fire hazard and for the development of methods to provide increased safety. In particular, these scenarios should provide the basis for materials selection, design criteria, validation of test methods, promulgation of specifications, and research and development objectives.

Conclusion: The application of fire scenario analysis appears to be a most productive method to identify effective means to improve fire safety in today's increasingly complex ships. *Recommendation:* Base the design and procurement of any new ship having polymeric materials, in which fire is a design consideration, upon the development and testing of the design against appropriate fire scenarios. This, in essence, requires the development of a fire impact statement.

Conclusion: At present, insufficient use is being made of the fire scenario technique in the analysis of shipboard fires. *Recommendation:* Train ship designers and operators and regulating authority personnel in the development and use of fire scenarios, to enable them to identify critical fire hazard elements and to determine appropriate protective measures.

2.4 Conclusions and Recommendations. Polymer Materials (Chapter 4)

Conclusion: Given sufficient oxygen and thermal energy input, all organic polymers will burn. *Recommendation:* Initiate programs to increase basic knowledge of the relationship between the chemical and physical properties of polymers, the fire dynamics parameters, and the way these relations are affected by aging.

Conclusion: Some usages of polymeric materials in ships can seriously augment the fire hazard. *Recommendation:* Concern exists about potential fire hazards associated with the rapidly increasing use of polymeric insulating foams in ship construction. Support approaches to improve the fire safety of the high-volume low-cost polymers.

Conclusion: Many synthetic organic polymers burn in a manner different than that of the more familiar natural polymers such as wood, paper, cotton, or wool. *Recommendation:* Increase the development effort on char-forming polymer systems, with particular emphasis on lowering the fabrication cost.

Conclusion: Poly-vinylchloride (PVC) and polyolefins have been and will remain in the near future the primary materials used in wire and cable insulation. Many factors, but primarily economics, dictate their continued use. *Recommendation:* Define problems, if any, associated with new PVC formulations. Develop approaches to surmount these difficulties. Develop comparative data on "fire-resistant improved" PVC formulation(s).

Conclusion: Fluorolefins will receive increased attention. They are limited to particular applications because of the high materials cost. There may be certain problems caused by the toxicity of combustion products associated with their use. *Recommendation:* Define and attack any problems associated with fluorine-containing polymers.

Conclusion: Polyphosphazenes are the only new more-fire-retardant materials currently approaching commercialization. They are emerging as a new class of higher priced materials to consider for use in fire retardant applications. *Recommendation:* Encourage future development of polyphosphazenes, defining advantages and problems.

Conclusion: Silicones and polyimides (primarily film) are two classes of commercial cable-insulating materials which do not depend on halogens (either as an addition or as part of the structure of the polymer) to provide fire-resistant properties. However, they do not fit into the same economic framework as PVC and polyethylene. *Recommendation:* Develop to the flame retarding of polyolefins by novel non-halogen approaches.

Conclusion: The fire safety of many polymers has been improved by the incorporation of hydrated alumina and/or compounds containing halogens, phosphorus, or antimony. *Recommendation:* Create an overall program which will categorize and communicate the goals and results of government-supported work on the fire safety of polymeric materials, particularly the use of additives.

Conclusion: The polymeric materials with improved flammability characteristics which are available today may have deficiencies such as high costs, fabrication difficulties, and formation of toxic and corrosive combustion products. *Recommendation:* Initiate programs to determine the relationship of chemical and physical components of polymeric materials to the formation and evolution of smoke and toxic gases.

Conclusion: The use of intumescent coatings can enhance the fire safety of some polymeric products, but may have limited usefulness in a marine environment because of adverse effects of moisture and salt water. *Recommendation:* Expand materials and application studies of intumescent coatings, with emphasis on lowering the cost and improving coating performance in a marine environment.

2.5 Conclusions and Recommendations. Test Methods, Specifications and Standards (Chapter 5)

Conclusion: The use of ASTM D-1692 test for flammability of furniture is not recognized as an acceptable test by the fire community. *Recommendation:* Consider the modified International Standards Organization (ISO) test method now under development to replace several of the Coast Guard flammability tests and as an eventual replacement for ASTM D-1692. In the interim, test fabrics by Federal Standard 191, Method 5903, and other polymeric materials by ASTM E-162, with appropriate limits of flame spread.

Conclusion: The testing of carpets by the ASTM E-84 test method is no longer considered a valid procedure for carpets. *Recommendation:* Use the new National

Bureau of Standards Flooring Radiant Panel test and the Federal Standards Test Method 372 in place of E-84 as a flammability specification for carpeting and floor coverings.

Conclusion: The Coast Guard has no flammability specifications for polymeric materials used in small boats, despite the particular susceptibility of pleasure craft to fires because of inexperienced crews and poorly safeguarded fuel systems. *Recommendation:* Consider using the modified ISO test method, after development is complete and appropriate limits have been set by test data, for polymeric materials used in small boat construction. In the interim, preparation should be made to monitor the development of the ISO method or other acceptable test methods to achieve a series of specifications for the various categories of materials.

Conclusion: Fuel systems are not a part of the Committee's charge; however, the design of fuel systems, in small boats clearly needs additional consideration to improve safety. Although the Coast Guard and Navy use tests that are presumably adequate, many of the test methods available to them are inadequate because they do not provide complete guidance for the selection of polymeric materials to be used in systems over a wide spectrum of fire scenarios. *Recommendation:* The Navy and the Coast Guard should intensify current efforts to monitor test method development, to adopt test methods in areas where none are currently specified, and to develop new test procedures.

Conclusion: The Navy specifications have no requirement limiting smoke emission from upholstery materials, which are frequently heavy contributors of smoke. In addition, the specification provides that flaming drippings may continue to burn for 5 seconds. *Recommendations:* The Navy should establish a smoke emission standard for upholstery in which the specific density (D) should not exceed 100 in less than 4 minutes by the National Fire Protection Association No. 258 test method. Change the specification to forbid flaming drippings.

2.6 Conclusions and Recommendations. Smoke and Toxicity (Chapter 6)

Conclusion: There is a lack of information on the special problem of combustion and pyrolysis, and the composition of smoke and other products from the combustion and/or pyrolysis of polymeric materials used aboard ships. The performance of a polymeric material under laboratory conditions may differ markedly from that which occurs when it, individually or in combination with other materials, is involved in a real shipboard fire. *Recommendation:* Develop a research program on smoke and toxicity problems that are specific to shipboard use. Every effort should be made to take advantage of actual shipboard fires to acquire information that can lead to improved specifications of polymers and enhance the medical care of victims.

2.7 Conclusions and Recommendations. Surface Ships — Commercial and Military (Chapter 7)

Conclusion: The fire safety record of Coast Guard-regulated vessels is good, relative to similar statistics on land transport or buildings. The fire safety regulations and specifications of polymeric materials in Coast Guard- and Navy-regulated vessels are much more stringent and closely controlled than similar regulations in land transportation or buildings. The absence of flammability standards for hull materials used in naval ships could lead to the uncontrolled use of polymers in the future. *Recommendation:* It is recommended that flammability regulations be set on naval ship hull materials.

Conclusion: The fire safety specifications regulating polyurethane foams in both commercial and naval ships are inadequate. *Recommendation:* It is recommended that the fire safety regulations controlling the use of polymeric foams in both naval and commercial ships be reevaluated and strict standards be set that rigorously control the design use of polyurethane foam.

Conclusion: The large amounts of sensitive electronic equipment on naval combat ships and their need to maintain full operational capabilities at all times make these vessels particularly sensitive to the corrosive gases produced by fires in most of the cable insulation currently in use. *Recommendation:* The current research programs to develop fire retardant electrical insulation which will not produce corrosive gases when burned should be expanded, and some interim fire protection standards that take account of corrosive combustion product should be promulgated.

Conclusion: The current fuel load calculation, based on the heat of combustion of wood, used to estimate the total fuel load in compartments, is misleading because it makes no provision for the much higher heats of combustion of polymers compared to wood. *Recommendation:* The present technique for calculating the fuel load in ship compartments should be modified to make allowance for the higher (relative to wood) heats of combustion of most common synthetic polymers.

Conclusion: The flammability and smoke specifications controlling the use of polymers in shipboard furnishings currently allow the use of many highly flammable and smoke-generating polymers. *Recommendation:* The flammability and smoke specifications controlling the use of polymers in shipboard furnishing should be reevaluated, preferably using the scenario analysis technique; more appropriate specifications should also be set.

Conclusion: The flammability regulations governing the use of polymers of insulation in small gauge communciations wire are probably unrealistic and inadequate because tests are run on individual wires, while in actual use many wires are invariably combined in large cables or cableways. *Recommendation:* The flammability regulations governing the use of polymers as insulation of small gauge communications wire should be reevaluated and reestablished, to better correlate with actual conditions.

Conclusion: The use of unprotected polymeric foams in the cargo spaces of commercial shipping allowed by present Coast Guard regulations, is a serious fire hazard. *Recommendation:* The flammability regulations allowing the use of unprotected polymeric foams in cargo spaces of commercial vessels should be re-evaluated, using the scenario analysis technique.

Conclusion: The use of fire resistant compartment penetrations is an effective method of reducing the fire spread along cableways. *Recommendation:* Effective fire resistant breaks for cableway penetrations in compartment bulkheads should be required on all naval and commercial ships.

Conclusion: The flammability specifications controlling the use of polymeric foams on the bulkheads and water piping of Navy ships may be inadequate because of the flammability test chosen. *Recommendation:* The flammability test specifications regulating the use of polymeric foams in the engineering and cargo spaces of naval vessels should be reevaluated, and toxicity specifications for pyrolytic combustion gases from these materials should be set.

Conclusion: The flammability and smoke specifications set in MIL-STD-1623 for light diffusers in naval ships can allow the use of some highly flammable polymeric materials. *Recommendation:* The MIL-STD-1623 flammability and smoke specifications on light diffusers should be reevaluated in view of presently available materials, and new specifications should be set.

Conclusion: The present naval research programs to develop new fire retardant elastomers, cable coatings, and cushioning materials with less corrosive and toxic combustion products are addressing one of the most serious and current fire hazards on naval vessels. *Recommendation:* The Navy research programs to develop new fire retardant elastomers, cable coatings, and cushioning materials should be continued and expanded.

Conclusion: As is more fully demonstrated in Chapter 3, scenario analysis is a powerful tool in materials selection, design, criteria, validation of test methods, the promulgation of specifications, and the choice of research and development objectives. *Recommendation:* Scenario analysis should be a required part of the design and construction regulations for all new ship construction.

Conclusion: The absence of toxicity standards of combustion gases in Navy regulations on materials leads to serious safety hazards on ships. *Recommendation:* Toxicity standards for combustion gases from polymeric materials should be set to diminish the hazard of the use of these materials on naval vessels.

Conclusion: The present flammability standards based on ASTM E-84 used for controlling materials on Navy ships may be inadequate. *Recommendation:* The use of ASTM E-84 as a primary flammability test controlling materials used on Navy ships should be reevaluated.

2.8 Conclusions and Recommendations. Special Surface Vessels (Hydrofoils, Surface Effect, etc.) (Chapter 8)

Conclusion: High-performance vessels require minimum weight to achieve specified performance polymeric materials offer great great help in meeting this requirement. However, most polymeric materials used aboard special high-performance vessels are easily ignitable and produce heavy volumes of toxic smoke. *Recommendation:* Known hazardous polymeric materials currently used in inhabited areas of dynamically supported vessels should be replaced with available improved materials.

Conclusion: Vessel crews may be involved in fires in which it is necessary to continue the operation of the vessel while fighting the fire. *Recommendation:* Vessel crews should be required to wear clothing fabricated from commercially available fire-resistant fibers (e.g., treated wool, treated cotton, treated polyesters, aramids, and phenolics) to provide increased protection against fire.

2.9 Conclusions and Recommendations. Surface Boats and Crafts (Chapter 9)

Conclusion: The casualty and property loss data accumulated by the U.S. Coast Guard for registered boats show a very low injury rate and small property loss from fires and explosions. Most of the reported fires occurred in boats with wood or fiberglass-reinforced hulls and inboard engines. *Recommendation:* The low incidence and consequences of fires aboard boats do not at this time appear to support the imposition of fire protection regulations that would add to the cost of recreational boats.

Conclusion: Current fiberglass-reinforced plastic technology is such that products with greatly improved fire safety performance are available at moderate price. The fire safety of interior surfaces and furnishings can be improved substantially by state-of-the-art technology. *Recommendation:* Promote the availability of materials with improved fire safety performance. Carefully monitor the Coast Guard casualty data in order to identify potential fire hazards requiring new safety standards. Substitute less flammable high-pressure phenolic laminates or coated aluminum panels for wood and wood veneer in bulkhead and overhead paneling where there is great danger of fire.

CHAPTER 3

DEVELOPMENT AND ANALYSIS OF FIRE SCENARIOS

3.1 Introduction

Each of the hundreds of fires that occur yearly on ships is the outcome of a chain of events. Some of these are the result of the development and growth of the fire; others are the consequence of human behavior and/or automatic protection devices. Obviously if one of the links in the chain of events can be removed or altered, the end result of the fire would be altered.

A useful method for analyzing the fire hazard in a particular situation, or from a given product, is to construct typical fire scenarios, (i.e., actual or generalized detailed descriptions of fire incidents). This allows identification of the critical stages in fire development and suggests opportunities for prevention and control methods.

Scenarios have maximum utility if they meet two conditions: 1) they represent accidents causing a significant fraction of the annual loss from fire; and 2) they provide sufficiently detailed information to permit useful analysis.

Statistical analysis of accident data is a guide to the frequency of occurrence of a scenario. Only very limited quantities of data, however, are currently available. While statistical information delineates the general types of fires which commonly lead to fatalities (or major property loss), it is necessary to prepare representative scenarios containing all the relevant details of the fire challenge and the human (or automated) response. To benefit fully from this information, more and better data on fires are needed.

It must also be recognized that statistics are not applicable to certain classes of problems. For example, there is the infrequent catastrophe of major proportions where a statistically valid sample of events is not available; technological change may occur so rapidly that the time lag between the introduction of a new material, product, or structure and the development of a statistically significant accident history may be unacceptable. Hence, experience and judgment are very important in developing meaningful scenarios.

A practical range of ship fire scenarios can describe only a small fraction of fire incidents that could possibly occur. It is necessary, therefore, that these scenarios treat relevant factors which affect fire development in a way permit generalization. In particular, scenarios based solely on actual incidents will be retrospective in nature and will be incapable of predicting the effects of new designs and new materials on fire safety unless extensions of the scenario can be applied to new situations.

The scenario concept is not a new tool in long-range planning (Zentner-1975). However, only recently has the scenario concept been applied to fire-safety program planning. In 1976, the National Bureau of Standards Center for Fire Research employed 5,040 different fire scenarios in developing a research plan to reduce the nation's fire losses in buildings. The National Fire Protection Association in its publication *Fire Journal* documents the chain of events in many actual fires which it investigates. These real-life fire scenarios have been employed as the basis for in-depth studies of fires in specific residences, (e.g., one- and two-family dwellings (National Fire Protection Association FR75-1, 1975), mobile homes (National Fire Protection Association FR75-2, 1975), and nursing homes (National Fire Protection Association, 1972). There seems to be little use made of the fire scenario approach to improve fire safety in the maritime community.

This chapter is primarily concerned with the development and analysis of fire scenarios. Also included is a summary of statistics on ship fires. In keeping with the focus of the study (i.e., on improving fire safety by modifying materials or using them better), the physical behavior of fire is emphasized and the behavior of human beings is deemphasized. Nevertheless, it is obvious that people may enter into the fire scenario by: (1) preventing the fire, (2) starting the fire, (3) detecting the fire, (4) extinguishing the fire, (5) escaping from the fire, or (6) being injured or killed by the fire. The human psychological and physiological characteristics involved are beyond the scope of this report.

3.2 Statistics on Ship Fires

Before the fire scenario approach is discussed, it is important to introduce information on where, when, and how fires start on ships, and on the resulting loss of life and property.

3.2.1 Navy's Annual Fire Losses

The Navy's Annual fire losses, *in all categories,* for the fiscal years 1971—75 are shown in Table 3.2-1. These figures represent a very small fraction of the annual U.S. fire deaths (11,800) and building losses ($3.4 billion). These losses also are a very small fraction of the value of Navy property, including ships, currently estimated at $60 billion to $200 billion. Table 3.2-2 indicates the losses from the major Navy surface ship fires in the period 1953—1975. On the average these losses account for approximately 20 fatalities and $20 million a year.

3.2.2 Commercial Vessel Losses

Annually the U.S. Coast Guard presents a statistical summary of commercial vessel casualties that were reported to and/or investigated by Coast Guard marine inspectors during the previous fiscal year. The Congress, the Coast Guard, and other regulating or insuring organizations have used the findings of these investigations to establish standards and determine the need for legislation to improve the protection of life and property at sea.

The master of a U.S. ship is required by law to report to the Officer in Charge, Marine Inspection, U.S. Coast Guard, whenever a casualty results in any of the following:

(a) Actual physical damage to property in excess of $1,500;

(b) Material damage affecting the seaworthiness or efficiency of a vessel;

(c) Stranding or grounding (with or without damage);

(d) Loss of life;

(e) Injury causing any person to remain incapacitated for a period in excess of 72 hours, except injury to harbor workers not resulting in death and not resulting from vessel casualty or vessel equipment casualty.

Table 3.2-1. Summary of the Navy's Fire Losses (Including Marine Corps) *

FISCAL YEAR	NUMBER OF FIRES	AMOUNT OF LOSS PROPERTY	OTHER	NUMBER OF INJURIES	NUMBER OF FATALITIES
1971	1799	$72,977,579	$540,442	184	25
1972	1535	18,110,795	100	242	3
1973	1909	93,693,650	740,235	271	40
1974	1829	54,632,196	0	251	24
1975	1783	88,124,713	52,037	170	22

Source — Federal Fire Council Report on Fire Losses Categories include: Aerospace Vehicles, Buildings and Contents, Forest, Grass and Tundra, Shipboard, and others

Every event involving a vessel or its personnel which meets any of the conditions of a reportable casualty is of concern to the Coast Guard. A number of reportable casualties are not investigated by the Coast Guard each year simply because they are not reported. The statistical summary in Tables 3.2-3 and 3.2-4 represents fire casualties to commercial vessels which meet the above fire criteria.

3.2.3 Recreation and Pleasure Boating Losses

Under the authority of the Federal Boat Safety Act of 1971, the Chief, Office of Boating Safety, has the responsibility to collect, analyze, and annually publish statistical information obtained from recreational boat numbering and casualty reporting systems. The report, *Boating Statistics CG-357,* lists accident data for recreation and pleasure boats (Table 3.2-5). In case of collision, accident, or other

Table 3.2-2. Major Navy Ship Fires

Year	(A) Aircraft Carriers	Hull Number	Cause	Loss ($M)	Fatalities
1953	Leyte	CV 32	Catapult Explosion	?.	37
1954	Bennington	CVS 20	Catapult Explosion	20.	103
1960	Constellation	CV 64	Construction Fire	130.	50
1965	Kitty Hawk	CV 63	Machinery Space Fire	1.	2
1965	Ranger	CV 61	?	1.5	1
1966	Oriskany	CV 34	Flares	10.	43
1967	Forrestal	CV 59	Flight Deck Fire-Ordance	20.	134
1969	Enterprise	CVN 65	Flight Deck Fire-Ordance	5.	27
1969	Lexington	CVT 16	Machinery Space Fire	1.6	0
1972	Forrestal	CV 59	Arson	10.1	0
1973	Saratoga	CV 60	Arson	5.1	0
1973	Kitty Hawk	CV 63	Lube Oil Fire	0.2	6
1974	Enterprise	CV 65	Vast Space Fire-Welding	3.7	0
1975	Kennedy	CV 67	Collision/Fire	1.	1
			TOTAL (14 FIRES)	209.2	404
	(B) Other Surface Ships				
1969	Avenge	MSO 423	Construction Fire	3.5*	0
1970	Goldsborough	DDG 20	Boiler Explosion	?.	2
1971	Roark	DE 1053	Lube Oil Fire	1.0	0
1971	Knox	DE 1052	Fuel Spill	0.1	0
1972	Newport News	CA 148	Gun Turret Explosion	1.5	21
1973	Force	MSO 445	Engine Room Fire	4.0*	0
1974	Enhance	MSO 447	Lube Oil Fire/Explosion	1.2	0
1974	Marathon	PG 89	Fuel Leak	1.0	0
1975	Belknap	DLG 26	Collision/Fire	213.	7
			TOTALS (9 FIRES)	225.3	30

*Total Loss

casualty involving a motorboat or other vessel, the operator must file a report if the occurrence resulted in:

a. loss of life
b. personal injury involving loss of consciousness requiring medical treatment, or resulting in incapacitation for 24 hours or more
c. property damage in excess of $100.

3.2.4 Summary

While fires do occur on ships, resulting in injures, deaths, and property damage, the problem is not acute when compared to fire-related injuries, death, and property damage in land transportation vehicles or buildings. Overall, the available statistics seem to indicate a good fire safety effort.

3.3 Fire Scenario Development

3.3.1 Guidelines for Development

This section is concerned with consideration of the important elements about a

Table 3.2-3. Statistical Summary of Casualties to Commercial Vessels

Fiscal year	Explosion and/or fire-structure equipment, all others+				
	1976++	1975	1974	1973	1972
Number of casualties	215	129	109	97	131
Number of vessels involved	219	133	121	103	136
Number of inspected vessels					39
involved	61	23	26	21	97
Number of uninspected vessels					
involved	158	110	95	82	
PRIMARY CAUSE					
Personnel fault:					
Pilots–State	0	--	--	0	--
Pilots–Federal	0	--	--	0	--
Licensed officer–document					
seaman	5	2	1	2	7
Unlicensed–undocumented					
persons	7	--	5	5	9
All others	13	8	7	7	18
Calculated risk	1	--	--	0	--
Restricted maneuvering room	6	1	--	0	--
Storms–adverse weather	0	--	2	1	--
Unusual currents	0	--	--	0	--
Sheer, suction, bank cushion	0	--	--	0	--
Depth of water less than					
expected	0	--	--	0	--
Failure of equipment	63	65	36	29	55
Unseaworthy–lack of main-					
tenance	0	--	--	2	1
Floating debris–submerged					
object	1	--	--	0	--
Inadequate tug assistance	0	--	--	0	--
Fault on part of other vessel					
or person	8	6	12	6	5
Unknown–insufficient infor-					
mation	112	51	58	51	41

(Continued)

Table 3.2-3 Statistical Summary of Casualties to Commercial Vessels (Continued)

Fiscal year	Explosion and/or fire-structure equipment, all others+				
	1976++	1975	1974	1973	1972
TYPE OF VESSEL					
Inspected vessels:					
Passenger and ferry-large	1	2	2	1	1
Passenger and ferry-small	15	6	9	3	8
Freight	24	8	7	4	13
Cargo barge	2	--	--	2	1
Tankships	9	5	3	4	6
Tank barge	6	--	3	5	8
Public	1	1	--	--	--
Miscellaneous	3	1	2	2	2
Uninspected Vessels:					
Fishing	69	54	32	40	45
Tugs	37	24	23	14	27
Foreign	18	7	4	2	6
Cargo	*	*	5	3	4
Miscellaneous	34	25	31	23	15
TIME OF DAY					
Daylight	124	73	55	48	88
Nightime	81	46	45	43	38
Twilight	10	10	9	6	5
ESTIMATED LOSSES					
Vessel	22151	9903	8380	11358	21316
Cargo	2081	362	1554	796	792
Property	4491	644	377	131	84
VESSELS TOTALLY LOST					
Inspected	7	1	5	--	2
Uninspected	52	50	33	42	46

+ Does not include explosion and/or fire - cargo, vessel's fuel, boilers, pressure vessel

++ 1 July 1975 to 30 Sept. 1976

* Not reported

(Concluded)

fire which belong in a scenario. It is recognized that virtually all real fire investigations are handicapped by the absence of trained observers, especially at the early stages of the fire, so frequently one must guess what happened from fragmentary evidence. Nevertheless, it is useful to indicate what information is desirable. In some cases, one may want to set up a simulation of a fire scenario: to determine

Table 3.2-4. Statistical Summary of Deaths/Injuries Due to a Vessel Casualty

Fiscal year	Explosion and/or fire-structure, equipment, all others+				
	1976++	1975	1974	1973	1972
Number of casualties	24	13	11	6	21
Number of inspected vessels involved	4	2	4	2	28
Number of uninspected vessels involved	20	11	7	5	36
Number of persons deceased/ injured	15/28	4/15	10/9	2/8	23/41
PRIMARY CAUSE					
Personnel fault:					
Pilots-State	--	--	--	--	--
Pilots-Federal	--	/1	--	--	--
Licensed officer-documented seaman	1	1/	--	--	1
Unlicensed-undocumented persons	3	--	--	--	5
All others	3	/1	2	--	6
Error in judgement-calculated risk	--	--	--	--	--
Restricted manuevering room	--	--	--	--	--
Storms-adverse weather	--	--	--	--	--
Unusual currents	--	--	--	--	--
Sheer, suction, bank cushion	--	--	--	--	--
Depth of water less than expected	--	--	--	--	--
Failure of equipment	6	3/11	4	--	4
Unseaworthy-lack of maintenance	--	--	2	--	--
Floating debris-unsubmerged object	--	--	--	--	--
Inadequate tug assistance	--	--	--	--	--
Fault on part of other vessel or person	1	/1	--	2	--
Unknown-insufficient information	10	/1	3	4	5
TYPE OF VESSEL INVOLVED					
Inspected vessels:					
Passenger and ferry-large	--	--	--	--	--
Passenger and ferry-small	--	--	/1	--	--
Freight	4/2	1/2	--	/1	/4
Cargo barge	--	--	--	--	--
Tankships	--	--	--	1/	/4
Tank barges	/2	--	1/1	--	1/4
Public	--	--	--	--	--
Miscellaneous	--	--	2/1	--	8/7
Uninspected vessels:					
Fishing	7/7	2/7	1/2	1/1	2/2
Tugs	1/3	1/	/1	/1	1/2
Foreign	2/4	/2	--	/5	/2
Miscellaneous	1/10	/4	6/3	--	11/16

+ Does not include explosion and/or fire-cargo, vessel's fuel, boilers, pressure vessel

++ 1 July 1975 to 30 Sept. 1976

These data also reflect a relatively safe fire record: approximately 11 deaths, 20 injuries and loss of $15 M per year from fire on commercial ships involving the structure, equipment, furnishings, etc. The primary cause of these losses is failure of equipment.

Table 3.2-5. Summary of Fire Accidents on Recreation & Pleasure Boats (U.S. Coast Guard, 1976)

Total Boating	Other Fire and Explosion (not incl. fuel)				
	1971	1972	1973	1974	1975
Accidents	49	47	101	85	70
Fatalities	2	4	4	5	0
Injuries	4	5	27	7	22
Amount of Damage (Dollars x 1,000)	1,767	315	869.1	647.7	7 58.9

Note: These reported fire related damages should be compared to 1975 estimates of approximately 9 million boats, 15 million operators and passengers with the value of the boats estimated at $5 billion.

whether or not what one thinks happened could really happen, to instrument the test fire, and obtain quantitative data on critical fire dynamic modifications and/or the effect of substituting different materials, etc. Complete knowledge of the relevant factors is essential.

3.3.1.1 Pre-Fire Situation

In general, important events in the fire scenario occur long before ignition. Frequently, decisions made during the planning, design, and building of the vessel will profoundly affect the events in the fire chain. Therefore, it is essential that adequate attention be directed towards the pre-fire situation since, in some instances, an optimum outcome will result from action taken long before the fire begins.

The first step in the development of the fire scenario should include the gathering of data such as governing regulations, plans and specifications, builder's and manufacturer's records, and inspection records. Attention should be directed towards the rationale for material selection, how and where the materials were used, and how materials were installed. Specifically, did the materials meet the applicable specifications, were they used properly, and were they installed correctly? These and similar pieces of information are essential to the completeness of any fire scenario.

3.3.1.2 Ignition Source

In the majority of cases, the ignition source is initially a desired combustion process which leads to the "unwanted" fire. Examples are the discarded cigarette, the welder's torch, the stovetop burner, etc. Another large class of ignition sources involves failures of electric circuits or equipment. Other special cases such as lightning, static sparks, and spontaneous combustion, also start fires.

Enough information about the ignition source is needed to characterize it quantitatively. This is because in many cases the ignition of the target fuel is marginal. For example, weld slag may fall on a pile of rubbish, but ultimately may self-extinguish. A blowtorch may burn the bulkhead decorative panel for a second but may only char it, etc. Unless we know the details of the characteristics of the ignition source, we cannot say when a fire will result.

The desired parameters and units of the ignition source are:

maximum temperature ($^\circ$C)
energy release rate (cal/sec or watts)
time of application to target (sec)
area in contact (cm^2)

On a more sophisticated level, in some cases we may need to know the details of the heat transfer from the source to the target, which may be some combination of conduction, convection, and radiation. The degree of air motion or turbulence may influence spontaneous ignition of a heated vapor rising from a surface. Access to oxygen is important; for example, a target immersed in hot combustion products may not ignite because oxygen is excluded by the heat source itself.

The most important single fact to recognize about a potential ignition source is that, for solid polymers which are not readily ignitable, a "strong" ignition source will generally ignite the target while a "weak" one will not. The "strength" of the source depends on the energy flux and on the time of application to the target, and sometimes simply on the product of these two. (See Vol. 4, Sec. 3.3.2).

3.3.1.3 Ignited Material

The first material to be ignited by the ignition source is generally important in the scenario. The question is, given an exposure to an ignition source, how does the probability of ignition depend on the properties of the target material? The characteristics of the target material are crucial in determining whether ignition occurs. Thus, a detailed description of the relevant target material properties is vital to the scenario.

If the target material is a flammable liquid, its ignitability will depend on whether it is in the form of a stationary pool, a foam, a mist, or a spray. If it is a stationary pool, its initial temperature is crucial. If this temperature is below the flash point, ignition will occur only after sufficient heating to bring a substantial portion of the liquid to the flash point. If the initial temperature is above the flash point, ignition of the fuel vapors above the pool will occur immediately and the pool easily sustains burning.

In most fire scenarios, the target material is solid. The ignitability of a solid depends not only on its chemical composition but also on the energy (including radiation) balance at the surface (including radiation), on its thickness and thermal properties, on its configuration, and on its orientation.

Under the heading of chemical composition of a target material, the following factors are especially relevant: 1) the basic material may contain small percentages of additives (e.g., fire retardants) or impurities which may have major effects on ignitability. 2) if the material is hygroscopic, like cotton, the initial moisture content will vary over a wide range depending on prefire humidity, with important influence on ignitability. 3) if the material contains several major constituents, (e.g., flexible polyvinyl chloride containing a large proportion of plasticizer), the ignitability depends on the more volatile constituent, in this case the plasticizer. 4) the target will frequently be composite in nature, consisting of an outer skin material and an underlying material, either of which may contribute to or retard ignitability.

The importance of the energy balance at the surface is shown by attempting to ignite a single piece of wood, e.g., a two-by-four. No self-sustained burning will result (unless the ignition source is applied for a very long time, so that the average temperature of the wood reaches about $320°C$). However, a match placed between two vertical two-by-fours close together will give self-sustained burning. The single piece of thick wood could not continue to burn because of the high rate of radiant energy lost from the charred hot surface to the cold surroundings. This effect is less important for materials which burn at lower surface temperatures, such as non-charring thermoplastics. Radiant input from the igniting source can also be important, so the reflectivity of the target material is also a significant factor in such a case.

The thickness and thermal properties of a material are vital in determining the time required to achieve ignition when a given heat flux is applied to the surface. This obviously becomes crucial in the scenario if the heat flux is of relatively short duration. A distinction must be made between "thermally thick" and "thermally thin" materials. (See Vol. 4 Sec. 3.3.2). The time to ignition for a "thermally thick" material is independent of the thickness and controlled by the "thermal inertia," the product of the thermal conductivity and the heat capacity per unit volume. For a "thermally thin" material, the time to ignition is proportional to the product of thickness and heat capacity per unit volume. (Fabrics are generally in this category). Whether the material behaves in a "thermally thick" or "thermally thin" manner depends not only on the thickness but also on the heating rate, the heating time, and the "thermal diffusivity," which is the ratio of thermal conductivity to heat capacity per unit volume.

In the case of a thin flammable material (deck covering, paneling, etc.) in thermal contact with an underlying material, the thermal properties of the underlying material can influence the ignitability by the degree to which the underlying material acts as a heat sink.

The configuration of the target material can also be of great importance. The foregoing discussion has implied a one-dimensional geometry. In reality, ignition tends to occur more readily in a crevice or fold or at an edge or corner, etc., than in the middle of a flat surface.

3.3.1.4 Flaming or Smoldering Combustion

Some combustible materials may burn either in a smoldering mode, like a cigarette, or in a flaming mode. Also a material may smolder for a certain length of time and then spontaneously burst into flame.

In general, only solids with very low thermal conductivity, such as porous solids, or thin solids not in contact with a heat sink, such as a suspended cotton thread or a free-standing piece of paper, can smolder. (A sofa cushion made of polyurethane foam elastomer under a synthetic fabric cover can burn in the smoldering mode). Smoldering is characterized by much lower spread rates than flaming combustion.

Smoldering is important in that: a) the smoke or gases produced may permit detection of the fire at an early stage; b) the pyrolysis products may be toxic; and c) a transition to flaming after a long period of smoldering may produce a very rapidly growing flaming fire, because of the preheating of fuel and accumulation of combustible gases which has occurred during the smoldering period.

It is known that the character of smoke produced in flaming combustion is different from smoldering combustion of the same fuel. Consequently, this must be taken into account when selecting smoke detectors.

Smoldering, (e.g. in a mattress), may continue for a very long time. Therefore, scenario analysis should consider the possibility of a long time lag between ignition and active flaming.

The burning of charcoal is generally referred to as a glowing combustion rather than smoldering. The importance to the fire scenario is that cellulosic materials, after the flaming combustion is finished, continue to glow for a substantial time as the residual charcoal is consumed. During this time, the possibility for a resurgence of the fire exists.

Also, when a gaseous extinguishing agent such as carbon dioxide or a halocarbon vapor is applied to a fire, it may stop the flaming combustion, but a smoldering combustion may continue (deep-seated fire) and after a time the extinguishing vapor may dissipate and the flame rekindle. Thus a "one-shot" gaseous extinguishing system may not assure protection unless the fire is held in check long enough for effective manual measures.

3.3.1.5 Fire Spread

General

Unless a person is wearing or sleeping on the originally ignited item, the fire is not apt to do much damage until it has grown by spreading some distance from the point of ignition. The rate of spread is very important in the scenario, because it defines the time after ignition when the fire reaches a dangerous size. The "dangerous size" may relate either to the rate of generation of toxic and smoky products or to the difficulty of extinguishment. The ability to detect, fight, or escape from the fire depends on the time for the fire to reach a dangerous size, and on the spread rate.

Fire may spread either from one contiguous fuel element to the next, or by jumping across a gap from the initially ignited material to a nearby combustible item. These two cases are discussed separately.

Fire Spread Over the Initially Ignited Material

The rate of flame spread over a solid surface in the horizontal or downward direction is often quite slow, sometimes as little as one inch per minute. However, if the material is "thermally thin" (see Sec. 3.3.1.3), or has been preheated by radiation or convection from hot combustion products, the flame can spread quite rapidly. If the fuel is so arranged that upward burning can occur, it will occur very rapidly and at a progressively accelerating rate. If the fuel is arrayed as a lining of a corridor or duct, with the air supply coming in at one end and the combustion products exiting at the other end, the fire will spread rapidly from air entrance toward air exit until it penetrates sufficiently far into the duct so that the oxygen is exhausted. It will then stop spreading until more oxygen becomes available, after which it will move downstream. Thus, in this situation, the effect of ventilation is controlling. Since most ship compartments have forced ventilation, this must be considered in fire spread. Indeed, for any fire burning in a compartment with limited air supply, the rate of spread will decrease as the air becomes vitiated by combustion. However, use of ventilating fans or other deliberate actions by fire-fighters to improve visibility by ventilating the fire will have an accelerarating effect on spread rate.

Fire Spread to Secondary Material

If one assumes that the originally burning material is separated by a gap from the nearest secondary combustible, and the flame does not impinge directly on this secondary material, the fire will die after the original material is consumed, unless it can somehow spread across the gap. The possible modes will be considered.

(a) The fire may radiate directly on the target. (b) The fire may convectively heat the ceiling and upper walls, which then radiate onto the target. (c) Hot smoky gases accumulating under the ceiling may radiate onto the target. (d) A combination of these may occur.

In any case, the effect of radiation is to preheat the secondary material until it pyrolyzes, emitting flammable vapors. At this point two possibilities exist. Either the secondary surface may ignite, or a sufficient concentration of a flammable vapor mixture is achieved so that the original flame may spread through this vapor cloud to the secondary material.

Other modes exist for fire spread across a gap. If the original burning object is a thermoplastic, it will melt, and burning droplets may fall and ignite secondary fuels that they encounter.

If the original burning object is mechanically weakened and falls over, the collapse can provide a means of fire spread. Similarly, mechanical action associated with an ineffective attempt to extinguish the fire may lead to its spread.

In the case of overheated grease in a galley pan fire, the burning grease may spatter and self-propelled droplets may spread the fire. It is possible that an ember from a burning object may be propelled several feet by pyrolysis gases and cause secondary ignition.

3.3.1.6 Evolution of Smoke and Toxic Gases

General

Smoke and toxic gases are important to the fire scenario in at least five ways. (1) they may provide a means of early detection of the fire; (2) they interfere with visibility and, thereafter, with escape or with fire fighting; (3) they have psychological and physiological effects on humans, including confusion, incapacitation, and death. In a majority of instances, deaths are a result of the toxic combustion products and not a result of the heat and flames from the fire; (4) smoke may be important in the fire spread process by virtue of its radiation emission or absorption; and (5) substantial property damage may be caused by smoke and/or corrosive combustion products.

Automatic Detection

In regard to automatic detection, the first consideration is the rate of smoke movement from the fire source to the detector. Under a no-fire condition, the air movement in a compartment is determined by existing forced convection for heating, air-conditioning, or odor-removal purposes; or, by open doors and hatches, along with air movement induced by external wind and ship motion; or, by free-convective motions driven by heat sources. For very small fires, the buoyancy effect of the fire heat will be negligible and the smoke will follow the existing air circulation paths. When the fire becomes larger than some critical size, the hot fire plume will rise to the compartment overhead and then flow under the overhead, creating an entirely new circulation path in the compartment. If, before the fire, the upper portion of the compartment is warmer than the lower portion, then temperature-induced stratification will exist and smoke may rise halfway up and then spread laterally. For early detection of fires, the foregoing factors are crucial in determining detector response and location.

The next consideration is the response characteristics of the automatic detector to the smoke. This depends on the time-dependent concentration and particle size of the smoke at the detector, the velocity of smoke past the detector, the orientation of the detector to the flow, the smoke entry characteristics of the detector chamber, the operating principle of the detector, and the sensitivity setting of the detector circuit, the battery voltage, etc. It is especially important to note that different combustibles, or the same combustible flaming or smoldering under different ventilation conditions, produce smoke of different particle size and detection characteristics. It is also known that smoke may "age" after it is formed, (i.e., agglomeration of smaller particles into larger ones will occur with consequent effect on ease of detection).

Visibility

The optical scattering properties of the smoke depend strongly on particle size as well as concentration, so the vision-obscuring aspects which interfere with escape or fire-fighting are strongly dependent on the type of combustible and mode of combustion. For example, incomplete burning of polystyrene or rubber produces large soot particles capable of obscuring vision even at low concentrations.

The lachrymatory effects of gases such as aldehydes or acids associated with the smoke particles have been shown to be important in interfering with vision.

The visibility at deck level will generally be much better than at higher levels in a ship's compartment, so the possibility of crawling to safety is important. The height at which exit signs should be located is thus relevant. If sprinklers operate, both the cooling and entrainment effects tend to bring the smoke closer to the floor. Also, fog which may result from the employment of sprinklers will interfere with vision.

Toxic Effects

Smoke and toxic gases are far more important than heat and flame as a cause of death in many fires, and carbon monoxide is the chief toxicant, according to present knowledge. However, other specific substances which may be present in the smoke, such as acrolein, hydrogen cyanide, hydrogen chloride, hydrogen fluoride, carbon monoxide, phosgene, etc., may be very important in certain cases and may exhibit synergistic effects. (See Volume 3 of this series for a more complete discussion.

The critical survivable concentration of toxicant depends on the time of exposure, which, when escape is not possible, depends on the history of the fire. Also, combined effects of toxicants with heat, excitement, loss of vision, etc., are believed important in determining survival, as is the original condition of health of the subject, and previous intake of alcohol or drugs.

Confused mental processes induced by toxicants may be of critical importance to survival in cases where the subject has to make a rapid and correct decision on proper escape tactics, or where the escape route is long and tortuous, as is often the case in large ships.

3.3.1.7 Extinguishment

At some point in the development of each scenario, either manual or automatic extinguishment activity may commence. This may involve smothering the fire or applying water or other agent. The techniques of extinguishment are outside the scope of this study. However, the effectiveness of extinguishment will depend on the burning characteristics of the polymeric combustible. If the fire has become too large or is growing too rapidly at the time extinguishment is attempted, the fire will not be controlled.

Accordingly the rate of fire spread and the maximum rate of burning of the fully involved combustible are important parameters. For manual firefighting, critical factors are how closely the firefighter can approach the fire, and how smoke prevents him from determining where the fire is. If he has a hand-held extinguisher of given capacity, will it be enough to do the job? When automatic extinguishers are present, there is generally no problem unless the fire is shielded from the extinguishing agent (i.e., in an unsprinklered locker) or unless it is a high-intensity fire, in which case the key variable is the extinguishing agent density.

3.3.1.8 Flashover

Flashover is a critical transition phase of a fire in a compartment. In general, it is important in a ventilated compartment, since otherwise the fire will tend to smother itself before the flashover stage is reached. Prior to flashover, when a local fire is burning in the compartment, the rate is determined by the extent of flame spread to that time. After flashover, all flammable contents in the compartment are burning or rapidly pyrolyzing, flames are projecting out the hatches, and the burning rate within the compartment is determined by the rate of ventilation and/or the total exposed fuel area. Flashover often occurs very suddenly, after an extended time of local burning, within a time interval of a few seconds, and is characterized by very rapid fire spread throughout the compartment, with flames violently rushing out the doors or other openings.

Whether flashover can occur in a compartment depends on its size and shape, the ventilation available, the intensity of the initial fire, and the quantity and disposition of secondary fuels. If flashover can occur, the time required for its occurrence will depend on the foregoing variables plus the thermal inertia of the compartment, especially that of the overhead area. A fire in a typically furnished room will require 5 to 20 minutes after flaming ignition to reach flashover; on ship, the factors of ventilation, compartment closure, etc., can markedly alter this time.

In the pre-flashover period, the upper portion of the room is filled with hot, smoky, oxygen-deficient gases. The lower portion contains relatively cool, clean air. At some intermediate level, perhaps two feet under the overhead, there may be both sufficient oxygen and sufficient heat so that target fuels at this height could readily ignite.

Radiation is probably of major importance in flashover. Thus, the infrared emission, absorption, and reflection characteristics of objects and smoke in the compartment are highly relevant.

The larger the volume of a compartment, the less likely it is that a fire of given size will cause flashover. Data on simulated room fires indicate that, for a 12' X 12' X 8' room, a fire consuming 2 pounds of fuel per minute could produce flashover in about 20 minutes, while if the combustion rate were doubled, i.e., 4 pounds per minute, flashover would occur in about 1.5 minutes. Thus, one would suspect that even a very large, sparsely furnished ship's compartment could flash over, if a high rate of initial burning is achieved, because the time to flashover is extremely sensitive to rate of heat release. The BOAC passenger terminal fire at J.F. Kennedy Airport, New York, in 1970 exemplifies a fire in a large space with low fuel loading but presumably very high heat release rate. This was the result of the burning of the plastic foam padding on lounge chairs: the padding apparently reached flashover condition in a short time. One can visualize an analogous shipboard situation in a crew's lounge or a cargo hold.

3.3.1.9 Spread to Adjacent Compartments and Catastrophic Failure

Fire-resistant compartmented ships are designed with the expectation that a fire in any one compartment will be confined by the structure itself so that either the fire is extinguished or the fuel is exhausted before the fire breaks through. Interior partitions, fire doors, etc., are subject to design constraints specifying that the partition can maintain its integrity for an appropriate time — 30 minutes, 1 hour, 2 hours, etc. — depending on circumstances. Therefore, the scenario should include information on the fire endurance rating of the relevant structural elements.

Assuming that the ship itself is fire resistant, the fuel loading, expressed as pounds per square foot, influences the duration of the fire, once it has grown large enough to become ventilation-controlled. As a rule of thumb, there is an endurance requirement of about 60 minutes for each pound per square foot of fire load. (This assumes certain typical ventilation rates). The fire load may range from a few pounds per square foot in lightly furnished occupancies to ten times higher in storage occupanices. The endurance requirements are based on the assumption that the fuel is primarily cellulosic; however, if it is primarily a polyolefin or rubber, the stoichiometric air requirement per pound of fuel will be as much as three times larger and the heat release per pound of fuel will be much greater. A ventilation-limited polyolefin fire may burn differently than a cellulose fire.

If the fire compartment has openings to other sections of the ship, such as open doorways, ventilating ducts, improperly firestopped or inadequately sealed openings in bulkheads, etc., these would become critical elements in the scenario.

Even if the fire itself is confined to the compartment of origin, the spread of smoke and toxic gases throughout the structure could have catastrophic effects.

If the fire is capable of heating structural elements of the ship to a failure point (e.g., steel, above $1000°F$), a collapse of the structure may occur. Thus, the thickness and integrity of insulation on structural elements becomes important to the fire scenario.

For multi-level ships, it is especially important to prevent any means by which fire may spread progressively upward from one level to the next higher. Key elements in this type of scenario are ventilation ducts, cableways, piping penetrations, fire endurance of the overhead, stairwells or elevator shafts, and improperly fitted closures, etc.

3.3.1.10 Essential Fire Scenario Elements

A scenario should cover as many as possible of the following points:

(a) The pre-fire situation.

(b) The source of the ignition energy should be identified and described in quantitative terms.

(c) The first material ignited should be identified and characterized as to chemical and physical properties.

(d) Other fuel materials that play a significant role in the growth of the fire should be identified and described.

(e) The path and mechanism of fire growth should be determined. Particular attention should be given to fuel element location and orientation, ventilation, compartmentation, and other factors that affect fire spread.

(f) The possible role of smoke and toxic gases in detection, fire spread, and casualty production should be determined.

(g) The possibility of smoldering combustion as a factor in the fire incident should be considered.

(h) The means of detection, the time of detection, and the state of the fire at the time of detection should be described.

(i) Defensive actions should be noted and their effect on the fire, on the occupants, and on other factors should be described.

(j) Interactions between the occupants of the ship and the fire should be detailed.

(k) The time sequence of events, from the first occurrence of the ignition energy flux to the final resolution of the fire incident, should be established.

The scenario should permit generalization from the particular incident described. It should provide a basis for exploration of alternative paths of fire initiation and growth and for analysis of the effect on fire safety performance of changes in materials, design, and operating procedures. When used in this way, the fire scenario can be an effective tool in increasing fire safety by increasing the capability to visualize and comprehend the events.

3.3.2 Capsule Scenarios

No two fires are alike in all details, however, all have certain elements in common, which permits a systematic study of fires and lead to generalized rules for increased fire safety. All fires have a cause; their initial growth is determined by the physical parameters of the system, and these parameters together with external control measures, intervene to limit the growth of the fire and determine the extent of loss. In this section we present a variety of brief scenarios, developed from real fire incidents in which polymeric materials played a significant role, to illustrate the diversity of ship fires and to show common factors that permit a scientific approach to fire safety. In preparing these illustrative scenarios, attempts were made to address the following areas: prefire environment, ignition source, material first ignited, other significant material involved, fire spread, method of detection, extinguishment, and extent of loss. Additional scenarios can be found in Chapters 7 through 10.

3.3.2.1 Conducted Heat Ignition Source

A ship was alongside a tender for upkeep and minor repairs. Damaged and worn-out plastic foam mattresses were being replaced. Several of these items were briefly stacked against a metal bulkhead until the hatch could be cleared for their removal. An engineering technician was using a torch to cut metal support clips from the opposite side of the bulkhead. He had properly requested, received, and instructed a fire watch on his side of the bulkhead. No fire watch was ordered for the other side because there were no openings in the bulkhead, the cutting torch would not penetrate, and the bulkhead was presumably devoid of combustibles. Yet, heat from the torch, transmitted through the metal bulkhead, ignited a mattress resting against it. The fire was discovered when smoke was seen pouring from an open hatch. Heavy black smoke hindered firefighting efforts. Fortunately, there were no combustibles other than the mattresses present to spread the fire. The fire was extinguished without serious structural damage, but with extensive smoke damage to the compartment of origin and adjoining compartment. Several crew members suffered from smoke and gas inhalation, but recovered without apparent permanent injury.

3.3.2.2 Upholstery Contribution to Flame Spread

Shortly after leaving a ship's lounge, a visitor on the vessel returned to the lounge to pick up an item he had forgotten. When he arrived, smoke was pouring through an open door. He immediately went to an alarm station and pulled the ship's general alarm. The lounge was provided with upholstered furniture, synthetic carpeting material, and vinyl wall covering. It is surmised that a cigarette caused ignition of a sofa which in turn ignited the carpet, subsequently spreading the fire to other furnishings. Rapid heat buildup resulted in flashover prior to arrival of the firefighting team. The bulkheads and decks, which were of noncombustible construction, limited the fire to a single compartment. The vessel's crew extinguished the fire. However, the lounge was a total loss. Firefighting efforts were hampered by extremely acrid smoke, Subsequent investigation indicated that, in order to meet fire retardancy requirements, the foamed plastic utilized in the furnishings had been manufactured with a halogenated fire retardant. When subjected to a high temperature environment, the fire retardant broke down, yielding, among other products, halogen acids.

3.3.2.3 Molten Metal Ignition of Trash

While a ship was undergoing repairs, a torch was being used to cut metal hangers from the overhead of a living compartment. Globules of molten metal and slag fell to the deck and down a nearby open hatch, landing in a trash container in the compartment below. Although a fire watch was present, this threat to the lower compartment was not noticed. The contents of the trash container included paper, discarded electrical and communications wire, wood scraps, and plastic packing material from components just installed. Ignition was followed by rapid growth of the fire in the trash container. Flames engulfed the nearby bunk mattress and spread to the wiring bundle in the overhead. The fire spread to nearby compartments through open doors and hatches before it could be extinguished. Large amounts of black smoke were generated and spread through the area, entering the exhaust side of the ventilation system. Smoke and fire damage were extensive in two compartments; substantial fire spread along the wiring bundles into other areas. Personnel were affected by smoke and gas; there were no fatalities.

3.3.2.4 Fire Spread by Electrical Insulation

A fire of unknown origin started in an unattended compartment. Supported by the combustible linings and furnishings in the compartment, the fire grew rapidly. Shipboard electric cables, incorporating relatively large quantities of various plastic and elastomeric compounds as insulation and jacketing components, were the primary means of fire propagation to other compartments. In addition, these polymeric materials generated dense black smoke and toxic and corrosive products of combustion which hindered the firefighting efforts. Damage from the fire was substantial, resulting in four months of unscheduled ship unavailability and $4.5 million in repair costs.

3.3.2.5 Summary

All of the foregoing fires can be characterized by the following generalized scenario:

a. Polymeric materials were deployed in a manner conducive to fire development.
b. An energy source was applied to an easily ignitable fuel element.
c. The fire grew and spread, consuming fuel and producing heat, smoke, and toxic products.
d. The fire was detected.
e. Fire control action was undertaken.
f. The fire was ultimately extinguished.
g. Loss resulted from the fire.

It is apparent that the fire could have been prevented or the loss minimized by more effective action at each step. Study of the details of the fire scenarios will identify the areas where the most effective measures can be taken to minimize fire impact in similar situations in the future.

3.4 Fire Scenario Analysis

3.4.1 Guidelines for Analysis

Prevention and control are the prime objectives of any fire scenario analysis, and a comprehensive, factual analysis rests heavily on the accuracy, level of detail, and completeness of the scenario. However, developing this kind of fire scenario requires either a completely documented report of a detailed post-fire investigation, and analysis specifically designed to determine how and where the fire started and progressed until extinguishment; a similar report of an instrumented full-scale test burn or a combination of both. In any case, until existing knowledge of the dynamics of actual fires is augmented by additional fire dynamics research, development of fire scenarios will be an art rather than a scientific discipline. Nevertheless, the application of fire scenario analysis appears to be a most productive methodology to identify economical, effective means to improve fire safety in our increasingly complex environment. It is especially applicable to ships.

The analysis of any fire scenario can be accomplished in various ways; one effective means would be to ask a series of pertinent questions concerning each essential fire scenario element. The answers to these questions should suggest means for prevention and control, while providing a basis for materials selection, design criteria, validation of test methods, promulgation of codes and standards, and research and development objectives. The following typical questions might be asked. The question can be used as a check list.

3.4.1.1 Pre-fire Situation

1. Were existing specifications met? If so, did they yield adequate performance? If not, why, and would they have been effective had they been enforced?
2. Were materials installed properly? If so, did they contribute to fire growth or did they help contain it? Would other installation procedures have been better?

3.4.1.2 Ignition Source

1. In as much detail as possible, what was the ignition source?
2. For how long was it in contact with the ignited material prior to flaming ignition? If this is not known, could it be determined by a separate experiment?
3. Could the ignition source be eliminated? How? By education? By design?

3.4.1.3 Ignited Material

1. What was the originally ignited material? If composite, what were the various layers?

2. What was the application of the material (e.g., mattress, insulation, cushion)?
3. How was it located relative to the overhead and nearest bulkhead? To other materials?
4. What were ventilation conditions in the compartment?
5. Did melting and dripping of the ignited material occur? Did this significantly affect the fire spread?
6. Did the ignited material collapse, fall over, or otherwise act to mechanically spread the fire? If so, what effect did this have on the events of the fire scenario?
7. Are there other materials that could have been employed in this application which would not have ignited under the same exposure conditions? If so, why weren't they used?
8. Do flammability tests on materials intended for this application adequately measure ignition resistance to level of the ignition source? Should they?

3.4.1.4 Flaming or Smoldering Combustion

1. Is it known if smoldering preceded flaming? For how long?
2. If unknown, was the ignited material capable of smoldering?
3. Can the volume (and composition) of gases produced by smoldering be estimated?
4. Can the time-dependent concentration of smoke and toxic gases arriving at a strategic location some distance from the fire be estimated?

3.4.1.5 Fire Spread

1. How long did it take for the first object to become fully ignited?
2. If flame spread to a second object, what was the mechanism of energy transfer?
3. Did one or two materials significantly control the fire spread rate? Could the substitution of different materials, or the incorporation of design modifications alter the rate of fire spread and growth?

3.4.1.6 Smoke and Toxic Gases

Automatic Detection

1. If a smoke detector was present, how was it located relative to the fire? Did it respond as expected?
2. If no smoke detector was present, how much sooner would the fire have been detected if a smoke detector had been present in a logical location? Would this have been soon enough to make a significant difference?

Visibility

1. Was visbility obscured for the firefighting party or in an escape route? When did this obscuration occur relative to detection time? Which materials seemed

to contribute significantly to visilbility obscuration?

Toxic Effects

1. Were victims affected by toxic substances?
2. What toxic substances caused death (autopsy)?
3. Did toxic substances interfere with victim's escape by promoting confusion or impaired decision making?
4. Could these toxic substances be attributed to any one material?
5. Did victims have pre-existing conditions such as limited mobility, circulatory disease, recent alcohol or drug intake, etc.?

3.4.1.7 Extinguishment

1. How large was the fire when first detected? What were visibility conditions at the time?
2. How much time elapsed between detection and attempted extinguishment?
3. How large was the fire when extinguishment was attempted?
4. What was the extinguishment technique and how successful was it?
5. If automatic extinguishment equipment had been present, how much sooner would it have been expected to control the fire, and how much less might the loss have been?

3.4.1.8 Flashover

1. Did flashover occur? How long after ignition? How long after detection?
2. Can essential elements in the fire growth and spread process be identified in relation to flashover?

3.4.1.9 Postflashover

1. Did fire spread beyond initial compartment? How? Was the door or hatch open?
2. How was ventilation system involved in fire spread?
3. Did bulkheads, fire doors, etc., fail? If so, after how long?
4. Did fire spread to deck above? By what mechanism?
5. Did structural collapse occur?

3.4.1.10 Summary

The availability of accurate, detailed, complete fire scenarios allows opportunity for an in-depth fire hazard analysis as illustrated by the foregoing of questions. As these questions are raised, and some answered, means for fire prevention and control will emerge. These may involve more education, better material selection, improved designs, installation of detection equipment, more stringent specifications, etc. However, whatever solution emerges, it will be based on a comprehensive systems analysis of the problem. This is the fire scenario analysis approach to the fire problem.

3.4.2 Analysis of Selected Scenarios

The specific goal of this section is to demonstrate that fire scenario development and analysis is a productive methodology for improving the selection and use of polymeric materials to increase fire safety in ships. Two actual ship fire scenarios will be developed and analyzed in accordance with sections 3.3.1 and 3.4.1.

Description: M/V "Cunard Ambassador" Fire, Sept. 12, 1974 (Boyce et. al., 1975).

Early in the morning, a fuel oil fitting on a diesel propulsion unit in the main machinery space failed. Failure caused fuel oil, heated to 180°F, to be sprayed on a hot manifold resulting in ignition of the fuel. Within seconds the main machinery space filled with smoke. The alarm was sounded on the bridge and the chief engineer went to the carbon dioxide storage room to actuate the onboard total flooding system. There was an interval of approximately 5 minutes between fire outbreak and activation of the system.

Directly above and adjacent to the main machinery space was an emergency hydraulic control room. The deck separating the two spaces was steel. The intense heat generated by the fire in the machinery space ignited combustibles and flammable liquids in the hydraulic control room. Fire fighting efforts were severely hindered. The main machinery space could not be entered to ensure that the fire was extinguished because of smoke and untenability of the room caused by the lack of oxygen. Smoke and heat were still being generated in increasing amounts.

The decision was made to close all fire doors and as many fire dampers as possible, and to secure the vessel for abandonment. The crew remained on board to continue fighting the fire. The fire had now spread to staterooms above and adjacent to the hydraulic room. By the use of water and fire-fighting foam, the fire was brought under control. During the evening the intensity of the fires had diminished and in the morning the crew's firefighting efforts were directed at extinguishing smoldering fires. By the end of the day all fires had been extinguished. Damage from the fire was approximately $10 million.

Analysis

This scenario will be analyzed following the procedure of section 3.4.1.

Pre-fire

The vessel was built in accordance with structural standards of the Department of Trade and Industries, United Kingdom. Structural fire protection was in accordance with the 1967 Fire Safety Amendments (Proposed Part H) to the 1960 Safety of Life at Sea Treaty. The principles that guided the development of these amendments were:

a. Minimization of the use of combustible materials.
b. Division of the vessel into zones by structural and thermal boundaries.
c. Provisions for early detection.
d. Ready availability of fire-fighting equipment.

e. Protection of means of escape or access for firefighting.

f. Separation of accommodation spaces from the remainder of the ship by structural and thermal boundaries.

g. Containment of the fire to the space of origin.

Detailed requirements concerning all aspects of these principles are contained in the Part H Amendment. If a vessel is equipped with an automatic sprinkler system, reductions are allowed in the amount of thermal insulation required between adjacent compartments. This vessel was not equipped with a sprinkler system. The construction materials were:

Bulkheads	— Asbestos cement panels.
Ceilings	— Perforated corrugated metal, backed with non-combustible fibrous glass insulation.
Structural Insulation	— Varying thickness of non-combustible mineral wool.
Decks	— ¼- to ½-inch stiffened steel, covered with magnesite deck covering.
Linings	— Low flame spread — melamine plastic laminate, vinyl fabrics and paint depending on location.
Furnishings	— Accommodations (staterooms) furniture — steel case padding — cotton mattress — cotton with innerspring draperies — fire resistant material

Ignition Source and Ignited Material

The fuel oil fitting had been installed to prevent rupture of the fuel line from pressure surges. Its failure sprayed hot fuel oil on a hot manifold, quickly igniting the fuel.

Fire Spread

The fire spread throughout the engine room and casing and ignited polymers in the hydraulic control room through a steel deck. Carbon dioxide (CO_2) temporarily extinguished the engine room fire, but no water was available to fight the fire in the hydraulic control room. The crew attempted to reach the fire pump in the distiller room, but the entrance was immediately adjacent to the hot engine room bulkhead and could not be used. Fire pumps in the distiller room and in the generator room were not damaged. The fire did spread to a single stateroom adjacent to the main engines. It appears that this fire burned itself out. The fire did not even approach the main vertical boundary. The control room was not damaged.

The hydraulic control room was located on the next deck. Corridors on both sides of the vessel were burned in the area of the hydraulic control room. It was here that the fire began to spread horizontally through openings. Staterooms adjacent to and immediately opposite the exit from the hydraulic control room were severely burned, as was an adjacent storage room. The fire was slow in spreading through fire resistant bulkheads.

On the deck above the hydraulic control room, the fire area increased in size, partly the result of open doors. The fire continued to spread on the next deck because of the large open shop area. Although the fire was very intense in the shop area, it did not spread to any more staterooms at that level; some overhead damage occurred near the exterior stateroom boundaries. The fire entered one stairwell, probably when a door was opened.

On the upper three decks, the large amounts of combustibles in the restaurant, main lounge, and casino burned freely. The most intense fires occurred in the large open areas (i.e., the main dining room, the night club area, and the commercial shop area). Although the fire gutted these areas, it was prevented from spreading horizontally by the fire doors. Examination of the contents of these sections of the vessel, in particular the night club facility, revealed a large installed teak dance floor which added to the intensity of the fire, many combustible furnishings (i.e., molded fiberglass reinforced plastic chairs and combustible table tops) were also identified.

Structural damage occurred in only one of the three main zones.

Extinguishment

It is believed that the installed CO_2 system successfully extinguished the fire in the main machinery space. While this could not be confirmed visually, there was a reduction in heat and smoke generation.

Of the three fire pumps on the vessel, number 1 and 2 fire pumps were located in the main and interconnected auxiliary machinery space on the port side. No remote control of these pumps was available so they were out of action.

The third fire pump was located in the distiller room just forward of the main machinery space. The emergency generator was several levels above the distiller room. Control wiring for this pump, however, ran through the main machinery space. The intense fire in this space destroyed the cables and thereby rendered this pump inoperative. Therefore, the fire pumps were not adequately isolated from each other.

A further problem of access to the third fire pump existed; the emergency access to this space was adjacent to an uninsulated bulkhead of the main machinery space. Attempts to enter this emergency access from above were abandoned because of intense heat and smoke.

The fire was eventually extinguished, more than a day after it started, by a combination of water, fire-fighting foam, and the fire burning itself out.

Smoke and Toxic Gases

Heat and smoke severely hampered firefighting efforts, but fortunately did not hinder escape efforts.

Summary

Analysis of this fire scenario resulted in the following findings and suggestions for minimizing a similar occurrence in the future.

1. A major finding from this fire is the adequacy, regarding public safety, of the ship construction standards to which the M/V Cunard Ambassador was built. By limiting combustible construction materials and separation of the vessel into divisions by thermal and structural boundaries, fires can be contained for periods sufficient to allow the safe isolation and evacuation of passengers from the fire areas. In this instance, active fire protection systems were rendered inoperative, and yet passengers could have comfortably remained in the fire zones adjacent to the center section of the vessel.

2. The severe property damage indicates the need for review of ship construction standards to reduce such loss.

3. The concept of separation of fire pumps requires not only physical separation of the pumps, but also of all controls and cables.

4. Access to auxiliary fire pumps should be remote from uninsulated bulkheads of compartments containing the main fire pump.

5. Main vertical fire stop zones effectively halt fire progresss if all openings are closed.

Description: SS C.V. Sea Witch June 2, 1973 (U.S. Coast Guard, 1976).

At about 0042 EDST, the outbound American cargo ship SS C.V. Sea Witch lost steering control in New York Harbor, veered out of the channel, and struck and penetrated the anchored Belgian Tankship SS Esso Brussels, which was loaded with crude oil. Thirty-one thousand barrels of oil from three ruptured tanks ignited on impact, and high flames englulfed both vessels in the area of the collision within minutes, igniting containers, exterior paint, and other combustible material.

Flames spread on the water at a rapid rate, affecting immediately the starboard side of the SS C.V. Sea Witch and the starboard side of the SS Esso Brussels. The mate and crew members of the SS Esso Brussels prepared the forward and after port lifeboats for lowering. The fire spread was so rapid that only the port after lifeboat was lowered into the water, where it was eventually engulfed in flames. Some of the crew jumped overboard to avoid the flames. Twenty-six survivors were rescued by tugs and boats in the area. The master of the SS Esso Brussels and 10 crew members died from drowning or burn-related injuries after abandoning ship; one crew member died on board, and one crew member remained missing.

The Chief Engineer and a few crew members on the SS C.V. Sea Witch attempted unsuccessfully to assist in firefighting on the after deck. Smoke below the after deck drove the boatswain and several crew members to the weather deck, where they took refuge with the remainder of the crew in the after deckhouse. Despite a caution by the mate to remain with the ship, seven crew members wearing life preservers jumped overboard; they were subsequently rescued.

The smoke and heat within the interior of the deckhouse increased as time passed and the crew eventually congregated on the upper deck as the main deck, boat deck, and cabin deck became untenable. About an hour after the collision, the crew of the fireboat raised two ladders to the upper deck rail and rescued the 30 members trapped there. It was not known at that time that the electrician was still at his assigned fire station in the emergency generator room. He was rescued about an hour later.

Damage to the ships and cargo amounted to about $23 million.

Analysis

This scenario will be analyzed following the procedure of section 3.4.1.

Pre-Fire

The SS C.V. Sea Witch was the first of three container ships incorporating automation features built by a U.S. shipbuilder at Bath, Me. for American Export Lines. The vessel, built to class under American Bureau of Shipping standards, meets a one-compartment standard of subdivision.

The SS Esso Brussels was built in Sweden in 1960 to American Bureau of Shipping Class standards as an A-1 oil carrier. Carrying capacity was approximately 340,000 U.S. barrels when loaded to 95 percent capacity.

Ignition Source and Ignited Material

The force of the impact generated sufficient heat to ignite the oil cargo of the SS Esso Brussels. The 31,000 barrels of spilled cargo provided a ready path for fire spread. The fire grew rapidly to a severe condition. This scenario principally analyzes the ship's ability to resist spread of the fire into compartments and to prevent catastrophic failure.

Spread to Adjacent Compartments

Cargo in containers on the weather deck of the SS C.V. Sea Witch was consumed or severely damaged by the fire. The shell frame of some of these containers and remnants of cargo were all that remained. Light exterior sheathing of these deck containers burned away, exposing their contents to the fire. Some wooden floors of the containers ignited and provided additional combustible material. Containers sheathed with plastic-laminated wooden sides offered little resistance to the spread of fire, and were consumed. Cargo in some holds was severely damaged. In other cargo holds, particularly on the starboard side, heat radiated through the hatch covers and the side shell containers stowed adjacent to these boundaries exhibited smolder damage. Also noted was a vertical progression of fire through a tier of containers without affecting adjacent containers.

Paint on the exterior of the after superstructure was burned. The engine spaces were free of fire, but both the engine spaces and interior of the after superstructure suffered heavy smoke damage.

Port lights, although crazed, were all intact. Interior accommodations, constructed of Marinite panels with metal furnishings, were discolored from heat and suffered fire damage in staterooms along the starboard side of the deckhouses; the damage extended several feet inboard. The remainder of the interior suffered light to heavy smoke damage. Draperies in the staterooms forward and to the starboard caught fire, but the fire did not spread.

On the SS Esso Brussels, flames from the burning oil on the water ignited exterior paint and exterior combustibles. The interior of both deckhouses was paneled with a pressed wood panel board, which offered little resistance to the spread of fire. The fire consumed nearly all combustibles on and above the weather deck level and caused severe fire damage to those areas of the after deckhouse subjected to the intense heat.

Summary

Analysis of the above real scenario resulted in the following conclusions and recommendations:

Conclusions

1. The ability of the crew of the SS C.V. Sea Witch to survive in the after deckhouse for a period of about 1 hour, and the electrician for about 2 hours in the emergency generator room, while the stern of the vessel was engulfed in flames, can be attributed to the excellent structural fire resistance of the interior paneling and furnishings of the vessel.

2. The deckhouse interior furnishings and construction of the SS Esso Brussels, primarily made of combustible materials, were almost completely consumed by the fire. The complete and rapid spread of the fire through the living spaces of both deckhouses emphasizes the absence of structural fire protection aboard this vessel. Although the departure of the crew members from the vessel in a lifeboat was not successful, had they sought shelter aboard, they probably would have died in the deckhouse fire. The hazards of combustible construction within the living accommodations of a tank vessel was clearly demonstrated in this incident.

3. The rapid spread of smoke through the engine room and afterquarters complicated the survival efforts of the crew of the SS C.V. Sea Witch. The early abandonment of the engine room because of the dense smoke could have been a significant factor under different circumstances in which positive steps to combat the fire or maneuver the vessel were undertaken by the crew.

4. The deck cargo containers on the side of the vessel were first affected by the flames and heat surrounding the SS C.V. Sea Witch. The fire accelerated as containers ruptured or were consumed during the fire, exposing additional fuel to the fire. The progress of the deck fire was uninhibited either by the small separation space between bays or the boundaries provided by the containers. The absence of any effective fire stops to separate the deck cargo mass — some 30 feet high, more

than 320 feet in length, and extending the width of the vessel — compounded the problem of fire containment.

5. The hull of both vessels remained intact after being subjected to intense heat. This substantiates the suitability of steel as a structural material, as did the extensive damage to the nonferrous fittings on the SS E sso Brussels and aluminum containers on deck of the SS C.V. Sea Witch.

Recommendations from Analysis of the Scenario:

1. Check the adequacy of existing container construction standards and the possible need for additional shipboard fire protection standards in view of the rapid spread of fire through the containers during the fire.

2. Conduct further studies to develop methods through which the spread of smoke within the interior of burning vessels can be prevented. Ventilation systems should be designed to provide manual or automatic means, not only to prevent the spread of fire but also the spread of smoke.

3.5 Conclusions and Recommendations

Conclusions: Fire statistics indicate that the fire problem on ships is not acute when compared to fire related injuries, deaths, and property damage in land transportation vehicles or buildings. Although each ship fire is different, the development and analysis of fire scenarios leads to the identification of common elements. This allows identification of the critical states in fire development, suggests opportunities for fire prevention, and directs attention towards various methods for control. *Recommendation:* Develop a wide spectrum of generalized ship fire scenarios, as outlined in this chapter, based on real or credible incidents. The specific fire dynamic elements in these scenarios, (e.g., rate of fire spread, rate of heat release, etc.) should be further quantified when necessary by information obtained from full scale experiments. The ultimate goal is to develop the capability of employing small-scale experiments that can be used to predict full-scale events.

Conclusion: Scenarios should be prepared so as to permit generalization from the particular incident described. A good scenario provides a basis for exploration of alternative paths of fire initiation and growth and for analysis of the effect on fire safety performance of changes in specifications, materials, and design. *Recommendation:* Use scenarios for the analysis of fire hazard and for the development of methods to provide increased safety. In particular, these scenarios should provide the basis for materials selection, design criteria, validation of test methods, promulgation of specifications, and preparing research and development objectives.

Conclusion: The application of fire scenario analysis appears to be a most productive methodology to identify effective means to improve fire safety in our increasingly complex ships. *Recommendation:* Base the design and procurement of any new ship using polymeric materials, in which fire is a design consideration, upon the development and testing of the design against appropriate fire scenarios. This in essence requires the development of a fire impact statement.

Conclusion: At present, insufficient use is being made of the fire scenario technique in the analysis of shipboard fires. *Recommendation:* Train ship designers and operators and regulating authority personnel in the development and use of fire scenarios, so as to enable them to identify critical fire hazard elements and to determine appropriate protective measures.

REFERENCES

Boyce, W. G., Kerlin, D. J. and Sheehan, D. F. Merchant Marine Technical Division, Office of Merchant Marine Safety, U.S. Coast Guard. Report on Fire Aboard M/V Cunard Ambassador (1975). Carhart, H. Presentation to Ships Panel on Fire Safety Aspects of Polymeric Materials, Naval Sea System Command Headquarters, Washington, D.C., January 6, 1977.

U.S. Coast Guard, Department of Transportation, Boating Statistics 1975, CG-357, 1 May 1976.

Proceedings of the Marine Safety Council, Department of Transportation, United States Coast Guard, Vol. 34, No. 3, March 1977, p. 58.

Ibid, Vol. 33, No. 1, January 1976, p. 8.

Ibid, Vol. 32, No. 1, January 1975, p. 8.

Ibid, Vol. 31, No. 1, January 1974, p. 18.

Ibid, Vol. 30, No. 1, January 1973, p. 11.

Department of Transportation, U.S. Coast Guard, "SS C.V. Sea Witch — SS Esso Brussels (Belgium) Collision and Fire in New York Harbor on 2 June 1973 With Loss of Life" Marine Casualty Report No. USCG/NTSB-Mar. 75-6, released 2 March 1976.

"Reducing the Nation's Fire Losses, the Research Plan." Center for Fire Research, National Bureau of Standards, January, 1976.

A Study of One- and Two-Family Dwelling Fires, NFPA No. FR75-1, National Fire Protection Association, Boston, Mass. (1975).

A Study of Mobile Home Fires, NFPA No. FR75-2, National Fire Protection Association, Boston, Mass. (1975).

Nursing Home Fires and Their Cures, NFPA No. Spp-17, National Fire Protection Association, Boston, Mass. 1972.

Zentner, R. "Scenarios in Forecasting" Chemical and Engineering News, Vol. 53, No. 40 (October 6, 1975), p. 22 and references cited therein. Washington, D.C., Jan. 6, 1977.

CHAPTER 4

POLYMERIC MATERIALS

4.1 Introduction

This chapter of the report is organized into nine sections. Sections 4.2 to 4.8 provide broad general information on the state-of-art of the fire safety aspects of the main classes of the natural and synthetic polymers as they are used in ships. Section 4.9 contains conclusions and recommendations. Statistics of the total use of polymers in any instance have not been compiled. In addition, the regulations with respect to use of polymers based on flammability vary from being essentially nonexistent for pleasure craft to rigid specifications for use in commercial vessels of U.S. Registry and in U.S. Naval vessels. The *U.S. Navy Habitability Guidance List of Acceptable Materials,* Revision D (1977), contains the most recent information concerning suitability of shipboard materials. A report on a workshop: *Flammability, Smoke, Toxicity and Corrosive Gases of Electrical Cable Materials (NMAB 342)* (National Academy of Sciences, Washington, D.C.: 1978), has been published. A detailed report on the state-of-the-art of the fire safety of polymeric materials is the subject of Volume 1 of this series.

The optimum selection of materials from a fire safety standpoint is a difficult task because of the many, sometimes conflicting, characteristics which must be considered before a proper decision can be reached. Materials selection is a primary way to reduce the threat of fire. Other important aspects of coping with fire hazards include general fire consciousness in system design, use of structural materials with improved fire safety characteristics, fire detection, and firefighting.

Competing fire safety requirements must be reconciled in considering polymeric materials with improved fire safety characteristics. For example, it might not be tolerable to have a decrease in ease of ignition or flame spread which also leads to an increase in the production of smoke or toxic combustion products.

In Volume 4 (Fire Dynamics and Scenarios) the lack of basic understanding and knowledge of the relationship between the chemical and physical properties of polymers and fire dynamics parameters (flame spread, ease of ignition, etc.) is discussed, as is how these relationships are affected by aging.

There is also a serious lack of knowledge concerning the relationship of smoke and toxic gas formation during combustion to the chemical and physical composition of polymeric substances.

Additionally there is a general need for better communication and wider dissemination of results within the fire research community. There is concern that there is no general access to potentially important technical information. The United States Fire Administration (U.S. Federal Emergency Services) is seriously addressing this important gap in communication.

4.1.1 Approaches to Improving the Fire Safety Characteristics of Polymeric Materials

No organic polymeric material can withstand intense and prolonged heat without degradation, even in the absence of oxygen. Given sufficient oxygen and energy input, *all* commercial polymeric materials will burn. Many synthetic polymeric materials burn differently than the more familiar natural ones. They may melt and drip and often give off dense and acrid smoke. Although some are less flammable than more familiar materials like wood or cotton, others burn with a more intense flame and resist conventional firefighting efforts. This tends to promote panic and can lead to damage and loss of life, which might have been avoided with a better understanding of the performance of such materials in a fire. Better understanding of the fire hazards presented by polymeric materials must be promoted through educational efforts (National Commission on Fire Prevention and Control, 1973; Tabor, 1975). Metals will also exhibit some undesirable characteristics under fire conditions, (e.g., aluminum will melt). There are several avenues open to reduce the fire hazard of polymeric materials. Among these are:

1. Development of new polymers whose fire safety characteristics are inherently better than those of the well-known materials. Although some materials in this class are available commercially now, most are expensive.

2. Improvement of the fire safety characteristics of available and generally lower cost materials by adding fire retardants. At present this approach is commercially the more important of the two. It may take the form of an application of a coating to the surface of the materials, or incorporation of a fire retardant into the bulk at some appropriate state of processing. The pros and cons of using fire retardants are discussed in Volume 1.

3. Combination of two or more materials in a way which utilizes the best properties of each, for example, utilizing steel plates on each side of a plywood slab.

Polymers may be fire retarded by introducing into the bulk of the material such fillers as alumina trihydrate or by introducing active compounds referred to as fire retardants. The former lower the fire load either by heat absorption or by providing an inert diluent for the fuel. The latter are usually halogen, phosphorus, nitrogen, antimony, or boron compounds, and may be used in synergistic combination. One must also distinguish between reactive and non-reactive retardants, according to whether they form covalent bonds with the polymer.

The important variables in polymer flammability are summarized in Figure 4.1. The chart shows that the useful maximum temperature for polymers is 190°C. A select few can perform satisfactorily for an extended period up to 250°C. For very limited exposure, temperatures of 300°C have been achieved. The burning of a polymer solid is essentially a three-stage process, consisting of a heating phase, a thermal pyrolytic phase, and finally, ignition. The behavior of a polymer during the initial or primary heating phase depends to a considerable extent upon the nature

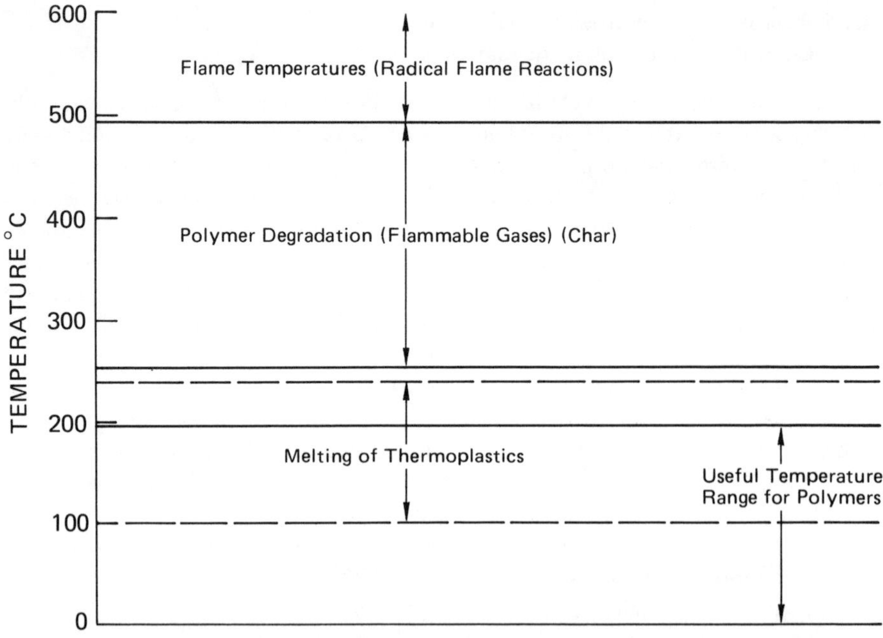

Figure 4.1. Polymer behavior at various temperatures.

of its composition. Thermoplastic compositions will generally melt in the temperature range of 100–250°C. The loss of rigidity that occurs at the softening point of such thermoplastic materials and the subsequent decrease in melt viscosity as the temperature increases allow the liquid phase to recede from the ignition source, and prevents the subsequent pyrolysis and ignition in many cases. This phenomenon has apparently led to some erroneous conclusions concerning the flammability of such compositions (Gouinlock, Porter and Hindersinn, 1971). Thermosets and most natural polymers such as wood and cellulose, on the other hand, remain essentially unchanged during this early heating stage.

At some later stage in the heating process, thermal decomposition occurs. Gaseous products emanate. The flammability of these products will depend upon the chemical composition of the original samples. The temperature and rate of heating determines when this stage occurs; this, in turn, depends upon the thermal stability of the material and chemical decomposition reactions occurring under the existing fire conditions. The flammability of a solid is largely determined by its behavior at this stage in the burning process. The establishment of a self-sustaining flame is predominantly dependent upon the generation of sufficient fuel gases from thermal pyrolysis to produce a flammable oxygen fuel mixture close enough to the solid fuel so that sufficient heat can be transferred from the flame to the solid surface by radiation and/or convection to sustain pyrolysis at an acceptable rate

(Essenhigh and Howard, 1966). This means that the flame zone is spatially removed from the fuel surface. This separation of flame from the solid fuel is necessary in order to allow dilution of the pyrolytic fuel gases with sufficient oxygen to bring the mixture within flammability limits.

Pyrolysis generally proceeds in three closely related phases. In the temperature range of 100–250°C, sufficient thermal energy is available only for such low-energy reactions as functional group elimination, usually from the end of the polymer chain, and the elimination of such small molecules as water and hydrogen halide. In the range 250–500°C, sufficient energy becomes available to break the highest energy chemical bond usually contained in the structure of most polymers.

These reactions can often lead to the unzipping of polymer chains to yield flammable monomers or the random elimination of small chemical fragments. Both types of products can sustain gas phase flame reactions. In some cases, however, recombination of some of these fragments also occurs and leads to the formation of aromatic condensed ring systems which are stable under the pyrolytic conditions. In these circumstances, a third stage of the pyrolysis occurs. Aromatic condensed structures formed in the previous stage are increasingly condensed at temperatures over 500°C, with the eventual elimination of most elements other than carbon. The result is a carbon char, which is highly insulating and flammable only with difficulty under normal oxygen concentrations. If the char can be maintained in a viscous elastic state during this intermediate stage, the gases will be trapped in the viscous liquid and thus cause the char to expand into a carbon foam. The formation of this special type of pyrolytic char is called intumescence. Such char forming reactions are thus desirable because they convert a flammable polymer to a less flammable char, while simultaneously reducing the quantity of flammable gases. If such a conversion can proceed, because of the nature of the polymer structure and in the absence of phosphorus, halogen, or heavy metal additives, highly toxic by-product gases are eliminated; the off gases are no more toxic than can be attributable to carbon dioxide or carbon monoxide.

With increasingly higher temperatures, the rate of production of the gaseous degradation products increases until a mixture is reached with the oxygen in the air that exceeds the flammability limit, and ignition occurs. Continued burning at this stage is dependent upon the transfer of sufficient heat from the flame to the condensed phase to maintain an adequate supply of flammable gaseous decomposition products and, of course, upon a supply of oxygen from the surrounding atmosphere sufficient to support combustion. The chemical reactions generally occurring in this gas phase at flame temperatures are free-radical in nature.

4.1.1.1 General Discussion of Fire Retardant Mechanisms

There are four main mechanisms or methods of altering the flammability of common commercial polymers. The first is the alteration or reduction of the heat of combustion of the total polymer composition. The second is the inhibition of the gas phase combustion reactions. The alteration of the condensed phase pyrolytic reactions to enhance the formation of char is the third most common method of changing the flammability of polymers. The fourth general method is the application of an intumescent coating. Exposure of the coating to the thermal flux of a fire expands the coating into thermally stable intumescent char which then protects a substrate from the ignition source.

4.1.1.2 The Heat of Combustion

It is generally conceded that the incorporation of an organic halogen compound into a polymer, either as an additive or by chemical reaction into polymer structure, will reduce the heat of combustion of the total polymer composition. An example of this effect is indicated in Table 4.1, in which the heat of combustion of various chlorinated methanes are compared to the unchlorinated precursor.

Table 4-1. Heat of Combustion of a Hydrocarbon and its Chlorinated Derivatives (Al'Shits and Flis, 1961).

Substance	Heat of Combustion (Kg Cal. per mole)
CH_4	−181.7
CH_3Cl	−135.1
CH_2Cl_2	− 72.8
$CHCl_3$	− 4.4
CCl_4	+ 47.5

Here we can see that successive substitutions in the methane series reduce the heat of combustion continuously until the nonflammable carbon tetrachloride exhibits an endothermic heat of combustion. Hindersinn and Wagner (1967) concluded that the heat of combustion was of secondary importance in polymer fire retardance because the fire retardant efficiency in the halogen levels was the reverse of the effect observed in the heat of combustion.

As additional support to their conclusions, these authors presented the data summarized in Table 4.2.

Here we can see that the highly flammable celluloid (nitrocellulose) has a heat of combustion of slightly more than 4 Kcal/g, or somewhat less than half of that exhibited by a fire retarded polyethylene composition. Further, the heat of combustion of a difficult-to-ignite polyvinyl chloride is about equivalent to that of highly flammable celluloid.

Gas Phase Inhibition

Gas phase combustion of hydrocarbon flames has been studied in considerable detail and many of the processes have been quantitatively defined (Fristrom, 1963).

*Table 4-2. Polymer Flammability Ratings and Heat of Combustion Values**

Material	Flammability	Heat of Combustion (KCal/g)
Celluloid	Very flammable	− 4.13
Polyethylene	Burns	−11.1
Polystyrene	Burns	− 9.6
Polyethylene (Fire Retarded)	Slow Burning	− 9.8
Polyester	Burns	− 4.3
Polyvinyl Chloride	Self-Extinguishing	− 4.3
Polymethyl Methacrylate	Burns	− 6.3

*(K. Krekeler & P.M. Klimke, 1965).

These reactions have been shown to be predominantly free radical in nature. The complete oxidation of a hydrocarbon can be explained on the basis of a complicated chain of free radical reactions. Initial attack on the saturated hydrocarbons is by hydroxyl radicals (generally the reaction product of oxygen and hydrogen atoms). In each step the products contained highly oxygenated products leading eventually to carbon dioxide and water. In this series, only reactions of hydroxyl and hydrogen atoms lead to an exponential increase in radical concentration. Inhibition of these latter reactions thus could disrupt the entire combustion chain.

Halogen Inhibition Reactions

The inhibition of these branching reactions by hydrogen halide has been suggested by many investigators as a major mechanism for the fire retardant effect of halogens on many polymeric materials. Halogen gas phase inhibition has been suggested as an explanation for the fact that certain bromocarbons were five to eight times more effective in reducing the flammability of hydrocarbon fuel gas mixtures on a molar basis than such inert agents as carbon dioxide and nitrogen. A mechanism for this flame inhibition has been proposed by Rosser, Wise and Miller (1959). It consists of the replacement of the radical chain carriers in the combustion series previously outlined by less reactive halogen atoms. The suggested mechanism of this type of halogen inhibition is summarized in Figure 4.2.

$$H\cdot \ + \ Hx \longrightarrow H_2 \ + \ X\cdot \quad \text{Inhibition Reactions}$$
$$OH\cdot \ + \ Hx \longrightarrow H_2O \ + \ X\cdot$$
$$X\cdot \ + \ RH \longrightarrow R\cdot \ + \ HX \quad \text{Regeneration of HX}$$

Figure 4-2. Halogen Inhibition Reactions.

In this series, the active hydroxyl and hydrogen radicals are converted by reaction with the hydrogen halides into water and the somewhat less active hydrogen molecules. The halogen atoms which are the other product in this series of reactions are much less active in oxidation reactions, but can react with the other hydrocarbons by extraction of the hydrogen radicals to regenerate the active hydrogen halide catalyst and produce a less reactive carbon radical. Although this mechanism was first proposed to explain the inhibitory effect of halogen compounds upon premixed hydrocarbon flames, its applicability to explain similar inhibitions in polymer diffusion flames is supported by the fact that the order of effectiveness of halogen in polymer fire retardants is the same as that observed in these premixed flames, (i.e., $Br > Cl > F$). In addition, it is well established that most, if not all, of the halogen in the polymer composition is liberated as hydrogen halide on exposure to a flame.

Condensed Phase Reactions

The large body of literature concerning the effectiveness of halogen inhibition of gas phase reactions has led to the postulate that it is the primary mechanism for polymer fire retardance. The gas phase reactions, however, are only the last in a complicated series of physical processes that must first occur before an adequate supply of flammable gases becomes available for sustaining an active flame as indicated in Figure 4-1. Thus, any change in condensed phase reactions leading to a reduced volume of flammable gases could lead to flame extinguishment, even in the absence of inhibition of the flame reactions. A careful analysis of the limited available literature concerning the effect of halogens upon condensed phase reactions indicates that alteration of these solid reactions is at least as important in many halogen/polymer fire retardant effects as the gas phase inhibition (Madorsky, 1964; Weiner, 1974).

From the above discussion of fire retardant mechanisms, it is seen that the gas phase mechanism is the least attractive from a fire safety point of view because the injection of inhibiting gases into the pyrolytic gas stream leads to incomplete combustion and large volumes of black smoke. Also, most of these inhibitors are toxic to a significant degree and many are in the highly toxic category. It should be noted that even though the introduction of fire retardant may reduce the probability of ignition it may also increase the optical density and toxicity of the smoke produced. A condensed phase mechanism, on the other hand, reduces flammability by conversion of the polymer to difficultly flammable char with a lesser production of smoke, and, generally, the production of pyrolytic off-gases no more toxic than carbon monoxide. Char-forming polymer compositions are, therefore, preferred from a fire safety point of view.

4.1.1.3 Economic Aspects

Cost is an important aspect of fire hazard reduction. Thermally stable polymers with superior fire safety characteristics may be too expensive for routine use. The application of fire retardant coatings is sometimes a cost-effective approach. However, this approach is limited to a relatively small number of applications.

It has been shown by Parker and co-workers (NASA-Ames R.C.) that one can produce polymers with fire safety characteristics that meet most end-use requirements by the synthesis of polymers with appropriate molecular structures. However, the high cost of these inherently fire retardant polymers must be reduced significantly if they are to gain large scale acceptance in most applications.

The cost of fire retardation by the incorporation of a fire retardant in the polymer varies greatly, according to the particular compound or treatment used, and according to the performance level desired. Fire retardation normally increases the cost of the material, except when the desired measure of protection can be obtained with inexpensive inert fillers.

4.2 Fibers

4.2.1 Introduction

Fibers form a main component of cushion covers, carpeting, draperies, and other ship compartment furnishings. Their flammability characteristics are thus of concern. Fibers will be discussed under the headings of natural fibers, synthetic fibers, glass fibers, and fibers from thermally stable polymers.

In Navy ships, the fibers are restricted to aromatic polyamides and phenolic/aromatic polyamide blends (aramid and novoloids) and/or fiberglass for draperies, curtains, and carpets. U.S. flag commercial vessels, designed for the purpose of engaging in trade, permit a broader range of fibers for these applications, (i.e., wool, nylon, acrylic, polyester blends, treated cotton, fiberglass, etc.).

4.2.2 Natural Fibers

4.2.2.1 Cotton

Cotton, which is essentially cellulose, is used as a base fabric for polyvinyl chloride coated-fabrics. It is used in bedding, (i.e., cotton sheets, in mattresses as cotton polyester blends and in draperies as treated cotton). Cellulose will burn under a wide variety of conditions. Cotton and rayon and their blends with thermoplastics (nylon, polyester, acrylic) are to be avoided because of ease of ignition and rapid burning rate. In principle, cotton in the form of fiber, yarn, or fabric can be treated with fire retardants in order to reduce flammability and thus could be acceptable for some applications. Fire retardants for cotton have been reviewed by Drake (1966, 1971). Successful and potentially acceptable durable fire retardants for cotton and rayon fabrics are of two general types, metal oxides and organophosphorus compounds.

Fire retardants based on metal oxides, especially in combination with halocarbon, have found greatest use in weather resistant textile products. Large quantities of these retardants are generally needed to impart sufficient fire resistance to a fabric to enable it to pass a vertical flame test. This, combined with the usually poor hand imparted by the halocarbons and loss during laundering, has limited the development of laundry-durable finishes of this type.

Organophosphorus fire-retardant compounds are made to penetrate the fiber where they react, polymerize, or copolymerize with an appropriate monomer or, in some systems, with the cellulose. Another way of applying organophosphorus compounds consists of depositing preformed phosphorus-containing polymers on the fibers or fabrics. Subsequently, these are either further polymerized or fused to provide durability.

Tetrakis(hydroxymethyl)phosphonium sulfate (THPS) is a fire retardant for medium and heavy weight cotton fabrics. Finishes based on (THPS), urea, methylol melamine, and various textile modifiers have been in use since about 1957, and are perhaps the most important methods for reducing the flammability of cotton. Other phosphorus compounds that have been suggested as fire retardants for cotton, but are much less important than THPC, are tris(1-aziridinyl) phosphine oxide (APO), reactive phosphonates such as N-methylol dialkyl/phosphono propionamide, phosphoric triamides in conjunction with polyfunctional N-methylol compounds, dialkoxy phosphinyl triazines, and diamides of alkyl phosphonic acids.

Generally speaking, good fire resistance is obtained on cotton fabrics through insolubilization of about 2 to 3 percent phosphorus, preferably in conjunction with nitrogen. Several excellent reviews have discussed the dozens of proposed experimental materials and the currently available knowledge on the mechanism of fire retardation in cellulose (Lyons, 1970; Kasem et al. 1972).

4.2.2.2 Wool

Wool is used in carpeting in all types of ships except for naval vessels. Wool textiles are generally more difficult to ignite, and spread flame more slowly than cellulosics. High concentrations of hydrogen cyanide have been found in the pyrolytic off-gases from wool products, and in some scenarios this may constitute a significant fire hazard. The flammability of wool has been reviewed in several recent articles (Benisek, 1972, 1972a, 1973; Friedman, 1973). Wool fire retardant methods have not been as extensively studied as those for cotton. Generally speaking, the flammability of wool is decreased by treatment with organophosphorus compounds, or by treatment with specific salts of polyvalent metals.

4.2.3 Synthetic Fibers

Synthetic fibers represent the "commodity" items of the textile industry and are produced in larger total volumes than cotton and wool combined. They are used extensively throughout ships. Such fibers include polyolefin, nylon, aromatic polyamides, polyester, polyester blend, acrylic, and inorganic fibers. With the exception of inorganic fibers, no fiber in this group can offer protection against direct exposure to flame. (See Volume 1, Chapter 3, for more detailed information on this subject).

4.2.3.1 Polyesters

The most important polyester fiber is poly(ethylene terephthalate). While poly-(ethylene terephthalate) is flammable, many polyester textiles will not ignite because the fabric melts away from a small ignition source or self-extinguishes by drip-out. Both of these mechanisms will fail with heavier constructions or when even a small amount of non-thermoplastic fiber (e.g., cotton) is present. Polyester produces an intensely hot flame when it burns, and therefore can ignite other adjacent materials.

To improve fire retardance in heavier weight constructions, it is possible to use several approaches. For example, the molecular composition of poly(ethylene terephthalate) can be altered to contain bromine e.g., using 2,5-dibromoterephthalic acid to replace some of the terephthalic acid. Bis(hydroxyethoxy)tetrabromobisphenol A may be used to replace some of the ethylene glycol in the ester molecule. Antimony oxide can be included in the formulation to enhance the fire retardance (Textile Industries, 1973). Bromine can also be introduced by means of an additive incorporated during the melt spinning process.

The comfort and aesthetics of poly(ethylene terephthalate) can be enhanced significantly by blending it with cellulose acetate or triacetate. When cellulose acetate contains appropriate fire retardants, or when fire retardants are topochemically applied to the blend fabric, good fire retardance can be obtained for some fabric constructions.

4.2.3.2 Acrylics and Modacrylics

Acrylics are polymeric fibers containing at least 85 percent polyacrylonitrile. Modacrylics (modified acrylics) are copolymers of acrylonitrile with vinyl chloride, vinylidene chloride and/or other vinyl monomers. Acrylic fibers must not be confused with acrylic plastics, which do not contain acrylonitrile.

Acrylics shrink away from a small ignition source and, therefore, escape ignition momentarily. Once ignited, however, acrylics burn vigorously, ejecting plumes of flaming gases and dense smoke. The rapid heat generation is due in part to a spontaneous "zipper-like," highly exothermic, cyclization reaction between adjacent nitrile groups. Acrylics are made acceptable for the "carpet pill test" by blending the face fiber with 20 percent of modacrylic fiber. Other attempts to improve flammability require copolymerization of acrylonitrile with a sufficient amount of other monomers to place the product among the modacrylics.

Modacrylics suffer from certain shortcomings in textile performance. Historically, they have a low softening point and poor thermal dimensional stability, although some improvements have been made.

The fire safety characteristics of modacrylics can be enhanced by copolymerizing with vinyl bromide. This does not impair the fiber properties, and may even improve them. The fire retardance is sufficiently good so that fabrics are being promoted that contain up to one-third polyester in conjunction with the modacrylic copolymer containing vinyl bromide. In one commercial application, antimony oxide also is present.

Allyl or vinyl esters of phosphoric or phosphonic acids are being offered as comonomers for incorporation into modacrylics, e.g., bis(2-chloroethyl) vinyl phosphonate. Addition of tricresyl phosphate or tris(dibromopropyl) phosphate to spinning solutions also are possibilities for improving fire retardance, but no commercial use of this concept is known to date.

4.2.3.3 Polyolefins

Polyolefin fibers are polyethylene or polypropylene. These fibers ignite readily and continue to burn with flaming drops. No formulas are known that successfully retard the flammability of polyolefin fibers. The use of retardants such as tris(dibromopropyl) phosphate and aliphatic chlorine compounds plus antimony oxide have been cited.

4.2.3.4 Nylon

Nylons are polyamides. The unmodified polyamides, nylon 6, nylon 66, nylon 610, etc., perform essentially like unmodified polyester with respect to flammability. Nylons resist ignition and flame propagation in vertical configurations because of drip-out, except in heavier weight fabrics. When nylon burns, however, it produces an intensely hot flame and can ignite adjacent materials. Carpets with nylon face fiber readily pass the methenamine fuel tablet test (see Volume 2) with even less face fiber density than poly(ethylene terephthalate). Nylon blends with cellulosics represent a flammability hazard. Published means for improving nylon fire retardance have serious deficiencies. Four basic approaches for obtaining nylon fabrics with reduced flammability have been cited (Stepniczka, 1973). Each method impairs some other desirable attribute (e.g., performance, durability or economics).

4.2.3.5 Aromatic Polyamides (Aramids)

This polymer system is the basis for the first commercially available thermally stable aromatic polyamide fiber. Draperies, curtains, and furniture upholstery are made of these materials for use as furnishings on naval shipboard.

Nomex® [poly(*m*-phenylene isophthalimide)] is duPont's first version of this class of polymers. Nomex® decomposes at about 370°C. Fabrics woven of the undyed material are more resistant to burning than solution dyed material (Ross, 1973). In low-flux thermal environments involving a small ignition source, Nomex® is self-extinguishing; however, when exposed to a large heat flux, it will shrink, burn, and propagate flame to other materials. As described below, improvements can be made by treating the woven or knitted form of the material. In comparison with other commercial fabrics, the Limiting Oxygen Index (LOI) of Nomex® (LOI = 27 to 28) is higher than most.

4.2.3.6 Fiber Blends

Blend fabrics made of yarns containing two or more fibers of different chemical composition and properties have attained great commercial importance in textile markets. But fiber blends pose synergistic fire hazards that often lead to unexpected flammability characteristics. For example, the inclusion of even small amounts of cotton into a polyester fiber garment can lead to a severe flammability hazard, because the non-fusible cotton prevents dripping of the fusible polyester, thus increasing the fuel available for burning. Early investigations have established that the fire safety aspects of blends cannot be predicted from a knowledge of the behavior of individual fiber components (Tesoro and Meiser, 1970). The physical and chemical interactions of different fibers in blends under conditions of burning pose complex problems which are not understood. Experiments on the flammability of blends have not been carried out during the last few years. Although fabrics have been developed from a wide variety of fiber blends, both synthetic and natural, polyester cellulose types have attained the greatest commercial utility.

4.2.4 Inorganic Fibers

For some end uses, inorganic fibers are mandated. Glass fibers, for example, melt at about 515°C but do not burn. They are generally treated with organic finishes to enhance their resistance to abrasion, and to improve other functional properties. The flammability hazard of glass fibers in actual use is thus assured by the presence of these organic materials. Although glass fabrics prepared in this manner are widely used as draperies, curtains, and other similar applications where non-flammability is necessary, they find their greatest utility as reinforcements for many different thermoplastic and thermoset resins.

Naval specifications permit glass fibers for carpeting, but modification of these specifications is under active discussion.

The thermal insulating characteristics, as well as fire resistance, are attributes of inorganic fibers and this has resulted in their widespread usage in shipboard applications. Asbestos, a natural inorganic fiber, has both fire and thermal resistance, and has been used extensively, until recently, in fiber, fabric, braid, batting and solid forms as thermal insulation (Kaswell 1963). Recent reports of the incidence of asbestosis and cancer in humans has resulted in a decrease in its use.

A synthetic inorganic fiber which is being used in place of asbestos is Fiberfrax®. This inorganic fiber is an alumina-silica composition (Kaswell, 1963), formed by blowing the molten alumina-silica. Because of the manner by which it is formed, its composition, and brittleness, Fiberfrax® can be made only as a staple fiber. The fiber has good thermal/insulating characteristics up to 2500°F. It can be formed into rovings, yarn, rope, braid, cloth, and tape (Floyd and Taylor, 1974); it is used for insulation, gasketing, packing, and filters.

4.3 Elastomers

4.3.1 Introduction

Most elastomers burn easily when not fire retarded. Presently, there is no elastomer that has the desired combination of low flammability, low smoke emission, good mechanical properties, and reasonable cost (Einhorn, 1971; Fabris and Sommer, 1973).

Elastomers are primarily used as electrical insulating materials for ships' power cables and communication cables. Ethylene-propylene copolymers, chlorosulfonated polyethylene, and silicones are the important elastomers for these applications. However, the Navy and Coast Guard are actively seeking materials with improved performance.

Since the fire retardation of natural and synthetic rubbers has been surveyed in detail by Fabris and Sommer (1973), this section is confined to a few general remarks.

The incorporation of halogens, either in an additive or as an integral part of the molecule, has been a prime approach to decreasing the flammability of elastomers. Thus we have polychloroprene (neoprene), chlorinated polyolefins, epichlorohydrin rubbers, the various fluoro and chlorofluoro elastomers, halogen-containing polyurethanes, and various compositions in which halogenated additives are used. All these materials are deficient in their fire safety characteristics (Gross et al., 1969): they give off smoke and hydrogen halides on combustion or exposure to an intense fire environment. In addition, some rubbers (particularly fluorinated materials) have the potential for generating other specific toxic combustion products.

Phosphorous compounds have also been used to decrease flammability. In the elastomer area, their use is generally limited to plasticized poly(vinyl chloride) and the polyurethanes. Materials that are fire retarded with phosphorus have somewhat slower flame propagation and are more difficult to ignite by small ignition sources, but they show little superiority when exposed to intense fire situations (Trexler, 1973).

A third approach to reducing flammability (and smoke) consists of replacing all or part of the carbon in the polymer structure with inorganic elements (Laur, 1970). The prime example of this is the family of silicone elastomers (Laur, 1970; Hooker Chemical Corp., 1970; Pepe, 1970) and the developmental phosphonitrilic elastomers (Hagnauer and Schneider, 1972).

A fourth approach consists in the incorporation of large amounts of inorganic fillers. This reduces the fuel value of the composition even if the filler has no specific fire retardant properties. Fortunately, most elastomers can tolerate and some even require a substantial amount (more than 50 percent) of particulate inorganic filler (Trexler, 1973).

4.3.2 Specific Elastomers

The fire safety aspects of elastomers are largely determined by their chemical structure. From this point of view, they may be conveniently assigned to several distinct groups.

4.3.2.1 Hydrocarbon-Based Elastomers

This group includes natural rubber, synthetic cis-polyisoprene, polybutadiene, styrene-butadiene rubber, butyl rubber, and ethylene-propylene rubber as main constituents (Kennedy and Tornquist, 1968; Bateman, 1963; Morton, 1973; Stern, 1967; Winspear, 1972; and Whitby et al. 1954).

These rubbers are low-cost materials with good mechanical properties, and thus are used in large volume for such applications as automobile and truck tires. They burn readily with production of smoke. Fire retardant additives reduce flame spread and ease of ignition from low-energy ignition soruces, but do not prevent burning in an intense fire situation. Alumina trihydrate is receiving intensive study as a filler to reduce flammability and smoke in these elastomers (Texas-U.S. Chemical Co., 1964, Hecker, 1968; Dalzell and Nulph, 1970; Hooker Chemical Co., 1970; Polsar, 1970).

4.3.2.2 Chlorine-Containing Elastomers

These include polychloroprene (neoprene), rubber hydrochloride, chlorinated ethylene polymers and copolymers (chlorinated polyolefins), and epichlorohydrin rubbers (Morton, 1973; Stern, 1967; Winspear, 1972; and Whitby et al., 1954). These materials have significantly better fire retardance than purely hydrocarbon rubber. They do, however, generate copious quantities of black smoke and hydrogen chloride gas when exposed to a fully developed fire.

Chlorinated elastomers, particularly polychloroprene are widely used where fire retardance is important. In ships, they are principally used in electrical insulation and in seat cushions and liners. The Navy specifies that chlorinated elastomers must be used in mattresses.

4.3.2.3 Polyurethane Elastomers

Polyurethanes are polymers containing the group $-NH-CO-O$ (Saunders and Frisch, 1962). They are formed typically through the reaction of a diisocyanate and a glycol. Because a variety of glycols or esters can be coupled with different diisocyanates, a large variety of linear polymers can be obtained in this way. These elastomers are crosslinked by including a controlled amount of a polyfunctional monomer (e.g., a triisocyanate or trihydric alcohol) in the reaction.

Fire retardant grades, generally based on bromine and/or phosphorus containing additives, are available. These, however, still burn in intense fires. Smoke generation is generally less than with hydrocarbon elastomers, but some hydrogen cyanide gas can be generated. The major use of polyurethane elastomers is in foams (seat cushions, insulation, etc.). These are discussed in Section 4.5.4.

4.3.2.4 Fluorocarbon Elastomers

As a class, fluorocarbons and fluorocarbon/ethylene copolymers are generally expensive. They are usually difficult to ignite and are not prone to propagate flames. Increased attention is being given to this class of materials. They do, however, have toxicity hazards from products of combustion and/or pyrolysis in intense fires.

4.3.2.5 Silicone Elastomers

Silicone elastomers generate relatively little smoke, are reasonably fire retardant in air, have low fuel value when burned, and do not contain halogen (Dow Corning Corp., 1969; Compton, 1967; Karstedt, 1970). They burn slowly and produce no flaming drip. They are relatively expensive (less so than fluorocarbons, but more than hydrocarbon rubbers), and for many applications their mechanical properties are marginal. Silicones do, however, offer one of the more promising combinations of fire safety aspects, physical properties, and cost, and are used primarily in electrical insulation on ships.

4.3.2.6 Phosphonitrilic Elastomers

Phosphonitrilic elastomers (Polyphosphazenes) represent a second example of "inorganic elastomers" (Hagnauer and Schneider, 1972). The phosphorus-nitrogen backbone:

$$\left[\begin{array}{c} \text{OR} \\ | \\ \text{P} = \text{N} \\ | \\ \text{OR} \end{array} \right]_x \quad \text{or} \quad \left[\begin{array}{c} R' \\ | \\ \text{P} = \text{N} \\ | \\ R' \end{array} \right]_x$$

supplies the flexibility required for elastomeric properties and contributes little fuel value. The various side groups (R or R') affect many of the characteristics of the elastomers, including their flammability. For example, long hydrocarbon side chains increase flammability while fluorocarbon side chains, although not contributing to flammability, contribute to undesirable pyrolysis products. These phosphonitrilic materials are in the early stages of development and much needs to be done to define their utility and feasibility for various uses. This includes identification of the combustion and pyrolysis products contributed by the phosphorus and nitrogen. Phosphonitrilic compounds are, however, the basis for hope for a low-smoke, low-flammability elastomer. The continued successful development of these new polymers will most likely result in their use as elastomeric foam and for electrical cable installations. The Navy is now evaluating these materials for fire retardant cable insulation.

4.4.1 Thermoplastic Resins

4.4.1.1 Polyolefins

The major polyolefins are low-density polyethylene, high-density polyethylene, polypropylene, and ethylene/propylene copolymers. Materials used in smaller quantity are polybutene, poly-4-methylpentene, ethylene/vinyl acetate and other copolymers and blends. Because of the low cost of the high-production polyolefins, economics will often dictate their use, despite some undesirable flammability characteristics.

Flammability Aspects

The combustion of polyolefins has been extensively reviewed by Cullis (1971). Chemically, polyolefins are very similar to paraffin wax. They ignite easily, burn with a smoky (less so than polystyrene) flame, and melt as they burn. The burning mechanism of polyolefins is similar to that for most solid materials.

The products of combustion of polyolefins generally are those expected from burning hydrocarbons; the major toxic material is carbon monoxide (Ball, 1973).

The use of additive systems based on a combination of halogen compounds and antimony oxide has been the most effective in reducing flammability. Although greatly improved resistance to ignition in low thermal energy environments can be obtained, all of these compositions burn readily in a fully developed fire, thereby contributing to the fuel load.

Crosslinked polyolefins are used extensively as cable insulation. They can be modified to improve their flammability characteristics in much the same manner as uncrosslinked polyolefins. Cable insulation can present serious flammability problems in spreading flame from one compartment to the next. IEEE 383 standards for "grouped" cables is providing an improved fire retardance test.

4.4.1.2 Styrene Polymers

ABS, one of the major types of styrene polymers, is finding increased use in thermo-formed hulls for small pleasure craft.

With the increasing amount and diversity of uses of foam and large thermo-formed items, relative hazard definition by use-analysis and meaningful testing is currently needed. Hazards arise when relatively large amounts of polymer are used and when large surface areas are exposed. In this way the hazard is enhanced by the high burning rate of polystyrene, and by the high temperatures and the dense smoke generated in polystyrene fires.

An extensive array of additives to reduce the flammability of styrene-based products has been tried (Lindemann, 1973; Howarth, 1973). The majority of these additives are halogen compounds, and usually incorporate a synergist such as antimony oxide. Generally, they are believed to function by increasing the depolymerization rate and promoting dripping of molten polymer, thus removing hot polymer from the burning sample. The beneficial use of additives has been emphasized in styrene-based foams.

4.4.1.3 Poly(vinyl-chloride) (PVC)

Vinyl chloride is an inexpensive monomer that can be polymerized by a variety of free radical catalysts to yield a high molecular weight polymer (PVC) with the general structure

$$\left[-CH_2 - \underset{\underset{Cl}{|}}{CH}-\right]_x$$

PVC itself does not burn under most normal conditions. The poor thermal stability of the polymer, however, generally necessitates compounding with significant and often large amounts of plasticizers or processing aids.

In ships, PVC is used in large amounts as cable insulation, in deck coverings and overlays such as in vinyl tile, vinyl coated fabrics (upholstery), as films on bulkhead, and in overhead sheathing (interior finishes), etc.

Flammability

When exposed to a flame or to excessive heat, PVC emits hydrogen chloride at relatively low temperatures in a highly endothermic process. This, together with the fact that the polymer contains more than 50 percent chlorine by weight, accounts for the low flammability of the uncompounded polymer. Depending upon the amount or type of the compounding ingredients added during fabrication, decomposition products may include, *inter alia,* benzene, hydrocarbons, and char. Large amounts of smoke can be produced. Despite these undesirable characteristics, the low cost of PVC will continue to dictate its use in many applications.

Chlorinated or phosphorus-based plasticizers are also used in large quantities to reduce the flammability of plasticized compositions. Among these, phosphates and chlorinated paraffins are used most widely. Phosphates, particularly tricresyl phosphate, cresyl diphenyl phosphate, and 2-ethylhexyl diphenyl phosphate, have long been added to PVC as plasticizers. They also enhance fire retardance and achieve excellent flame-out times. Many of the plasticizers or processing aids used in PVC are flammable, particularly the widely used phthalate, sebacate, and adipate esters, and various low molecular weight adipate polyesters.

4.4.1.4 Acrylics

Polyacrylates as thermoplastics and coatings are not used extensively in ships. Perhaps one of the largest uses for acrylics in these markets is as a decorative paint or coating. Polymethylmethacrylate is used in light diffusers and as glazing because of its good transmissivity, relatively high impact resistance, and resistance to degradation by light.

The acrylics are polymers formed from acrylic (R = H) or methacrylic (R = CH_3) esters according to the formula

$$\left[-CH_2 - \overset{\displaystyle R}{\underset{\displaystyle COOR'}{\overset{\displaystyle |}{\underset{\displaystyle |}{C}}}} - \right]_x$$

where R' represents an alkyl radical. The major plastic in this group is the homopolymer of methyl methacrylate ($R' = CH_3$), a crystal clear material that softens at about $100°C$.

Poly(methyl methacrylate) (PMMA) ignites readily and softens as it burns. Burning rate, fuel load, and smoke production are less than for polystyrene (Hilado, 1969). In burning, PMMA undergoes depolymerization from the heat of the ignition source, the heat of combustion, or other environmental energy (Conley, 1970), The volatile products of pyrolysis then burn in the gas phase.

Halogen and antimony compounds have been used to reduce burning rates and ease of ignition (Howarth, 1973). Less effort has been devoted to the fire retardation of PMMA than to that of other polymers, partly because of the realization that it is difficult to retard the characteristic "unzipping" depolymerization mechanism for most applications. Further, fire retarding additives usually detract from the excellent transparency and aging characteristics of the polymer. When used as glazing, particularly in relatively large areas, the potential fire hazard should be carefully analyzed.

4.4.1.5 Polyacetals

Little success has been achieved in making polyacetals more fire retardant because of the nature of their pyrolysis and their pyrolysis products. On the other hand, many of these polymers are used in relatively small parts where they do not present major fire hazards.

The commercial polyacetals are formaldehyde polymers and copolymers terminated (capped) with ester or other groups for stabilization. The simplest polyacetal is poly(methylene oxide)

$$\left[-CH_2 - O - \right]_x$$

Three principal flammability characteristics of formaldehyde polymers and copolymers are: low oxygen requirement for combustion (low limiting oxygen index), very low smoke production, and low fuel value.

4.4.1.6 Polyesters

The polyesters discussed in this section are linear thermoplastic poly-(ethylene terephthalate) (PET), poly(tetramethylene terephthalate) (PTMT) and their modifications. The crosslinked styrenated polyesters will be discussed in Section 4.4.2.3.

Thermoplastic polymers burn with a smoky flame, accompanied by melting and dripping and little char formation. Fire retarded grades are generally prepared by incorporating halogen-containing materials as part of the polymer molecules or as additives. Metal oxide synergists are frequently included. These fire retarded systems are resistant to small ignition sources in low heat flux environments but will still burn readily in fully developed fires.

4.4.1.7 Polycarbonates

Polycarbonates are a class of polyesters derived from bisphenols and phosgene. The commercial products are based on bisphenol-A.

Commercial unmodified polycarbonates are significantly less flammable than unmodified styrene, olefin, or acrylic polymers. On pyrolysis or burning, they produce some char. They extinguish during simple horizontal burning tests, and have an oxygen index significantly above those of the previously discussed unmodified thermoplastics (Hilado, 1969). Their fire resistance has been further improved by the use of halogenated bisphenols in the preparation of the polymer or by the use of halogen-containing additives with or without antimony oxide (Howarth, 1973). Recently several patents have been issued which describe the use of small amounts of perfluoroalkane sulfonates and aryl sulfonates as excellent flame retardants for polycarbonates in the absence of halogenated compounds (Nouvertine, 1973; Mark, 1975). Polycarbonates will burn in the presence of a high thermal flux. Little information is available on the toxicity of their combustion products.

4.4.1.8 Chlorinated and Chlorosulfonated Polyethylene (Hypalon)

Chlorinated and Chlorasulfonated polyethylene are used in relatively low volume in ships. Hypalon is used as fire resistant electrical wire and cable insulation.

Polyethylene can be chlorinated in the presence of light or free radical catalyst to give a chlorinated polymer of the general structure (Canterino, 1967).

$$\left[(CH_2 - CH_2)_x - (CH_2 - \underset{\underset{Cl}{|}}{CH})_y - (CH_2CCl_2)_z \right]_n$$

The chlorine content, and therefore the properties of the product, can vary considerably depending upon the extent of chlorination and the reaction conditions. The flammability decreases directly with the chlorine content. Various compositions are reported to extinguish under ASTM D-635 conditions. These are reported to contain 25 to 40 percent chlorine by weight. Compositions containing as much as 67 percent chlorine have been prepared. As with chlorinated polymers in general, antimony oxide enhances the efficiency of the halogen and reduces the amount of chlorine required to yield the desired fire retardant properties.

The flammability characteristics of these materials resemble those of poly(vinyl chloride). Hydrogen chloride is a major combustion product.

Polyethylene can be chlorosulfonated by methods similar to those used in the chlorination already described to give polymers of the general structure

$$\left[-CH-CH_2-CH_2-\overset{\displaystyle H}{\underset{\displaystyle SO_2Cl}{C}}- \right]_x$$
with the first carbon bearing Cl

The properties of the composition can be varied widely depending upon the extent of chlorosulfonation.

As expected, the flammability of the polyethylene is reduced as the chlorosulfonyl and chlorine content is increased. The flammability has not been studied extensively, although hydrogen chloride, sulfur dioxide, and considerable smoke are major products of combustion.

4.4.2 Thermosetting Resins

Thermoset polymers are distinguished from the thermoplastics discussed in Section 4.4.1 in that they become chemically crosslinked during the final molding. The final product is "set" into shape during formation of primary chemical bonds. For most practical purposes, a thermoset polymer can no longer be melted, shaped, or dissolved. Thermoset reinforced polyesters are used extensively in the fabrication of the hulls of pleasure craft and are discussed more extensively in Section 4.4.2.3. Generally, the polymerization of thermoset resins is divided into two stages. In the first stage, a relatively low molecular weight prepolymer is formed, which can be melted, dissolved or molded. In the second stage, the prepolymer is crosslinked (cured), with or without the application of heat in the presence of suitable catalysts, activators, or crosslinking monomers. Sometimes an inhibitor is added to the first stage to prevent premature curing.

Because of their brittleness, thermosets are used almost exclusively in conjunction with various inorganic or organic fibrous reinforcments and various types of powdered fillers. In many cases, the reinforcements and fillers comprise more than half of the final composition and can significantly alter the flammability or fire safety of the total composite. The total composition must be considered before deciding upon the flammability characteristics or fire safety of these materials.

Unlike many thermoplastic materials, thermosets generally do not soften or drip when heated. This is because of their crosslinked nature. Flammability is a function of the thermal stability of the primary chemical bonds and the ease with which volatile gaseous products can be produced by pyrolytic processes to provide fuel for a self-sustaining fire. Many thermosets (e.g., the phenolic resins) produce very little flammable fuel when heated by an ignition source, but degrade into an insulating char which can only be oxidized at extremely high temperatures and/or high oxygen concentrations. Burning of such materials can be a slow process under many conditions, since the polymer substrate is protected by the surface char. Such resins are inherently fire retardant and will pass many common laboratory tests without the need of a fire retardant modification or additive. Their fire retardance, however, is a function of the mechanical stability of the insulating char and is limited by the resistance of elemental carbon to oxidation.

The thermoset resins are in general more difficult than thermoplastics to mold and fabricate. This frequently militates against their use in many applications where their low flammability makes them desirable. Development of these high char forming materials to improve their ease of formation into molded parts at reduced fabrication cost should be vigorously pursued.

4.4.2.1 Phenolic Resins and Molding Compounds

Phenolic resins of various types find utility in ships but mostly in small moldings where flammability is not a factor. Two general types of first stage phenolic resins are produced, depending on the catalyst, the phenol/formaldehyde ratio, and the reaction conditions. These are called resoles and novolaks, respectively. The physical properties of the phenolic resins vary widely depending upon the type, kind, and amount of filler; the kind of reinforcement; phenol/formaldehyde ratio; type of curing catalyst; and other formulation variables. More information may be found in Brydsen (1970), Billmeyer (1971), and Foy (1969).

If a phenol/formaldehyde ratio of one to one or less is used in the initial condensation with an acidic catalyst, thermally stable resinous thermoplastic polymers can be formed which are commonly referred to as novolaks. These resins can be formulated subsequently with fillers, colorants, reinforcing agents, catalysts, and additional formaldehyde (or formaldehyde generator) to form the commercially important molding compounds.

Cured phenolic resins do not ignite easily because of their high thermal stability and great charring tendency in the pressure of heat (Sunshine, 1973). The flammability of the end products can vary widely, as mentioned above, depending upon the amount and type of filler used, the crosslink density, the amount and type of reinforcement, and other minor formulation variables. The principal volatile decomposition products are methane, acetone, carbon monoxide, propanol, and propane. A variety of additives have been found to be useful in applications in which a degree of fire retardance above that inherent in the polymer is required.

4.4.2.2 Unsaturated Polyester Resins

Because of their versatility and low cost, glass reinforced polyesters have found large volume usage in the hulls of small boats and pleasure craft. Marine applications accounted for about 25 percent of the total U.S. volume in 1975 and 1976.

Fire retardant compositions constitute only a small part (approximately 5.5 percent in 1973) of the total production of 550,000 metric tons in 1974 (see also Nametz, 1967; Roberts et al., 1964).

There are two classes of thermoset resins which are commonly referred to as polyester resins. These are the alkyds and the so-called unsaturated polyester resins. The latter are prepared by condensing a saturated dibasic alcohol and a mixture of saturated and unsaturated dicarboxylic acids into a prepolymer (or first stage) resin. The latter is then dissolved in a vinyl monomer, usually styrene. The cured resin is produced by free radical copolymerization of the styrene monomers and the unsaturated acid residues.

Phthalic anhydride is used most widely as the saturated acid component. The resins are usually combined with a reinforcing fabric (generally glass cloth or mat) and/or a filler before curing.

The burning characteristics of unsaturated polyesters can be modified by the addition of inorganic fillers; the addition of organic fire retardants; chemical modification of the acid, alcohol, or unsaturated monomer component; and by the chemical combination of organometallic compounds with the resin. A wide variation in flammability characteristics can be achieved in polyester resins by using one or more of these retardant modifications.

Both fire retarded and unretarded polyester resin formulations yield copious amounts of smoke when exposed to fire because styrene is the major product of pyrolytic decomposition and styrene burns with a very smoky flame. That smokiness has been only marginally reduced to date by the use of relatively large amounts of inorganic fillers such as alumina trihydrate.

The relative toxicity of the combustion products of halogenated polyester resins has been a subject of considerable discussion ever since their introduction in 1953. Generally, most, if not all, of the chlorine contained in these compositions converts into hydrogen chloride. Trace amounts of phosgene have been identified as well.

4.4.2.3 Epoxy Resins

Epoxy resins are generally prepared by reacting a first-stage polyfunctional epoxy compound or resin with a basic or acidic crosslinker (or "Hardener") to yield a thermoset product crosslinked by ether or ester linkages. The basic epoxy resin can be prepared in a variety of ways, although the most common is the reaction of a polyphenolic compound with epichlorohydrin.

The prepolymer can be cured with a variety of crosslinking agents (often incorrectly called catalysts) through the epoxy and hydroxyl groups. These hardeners can be based on amines, anhydrides, or Lewis acids.

Epoxies are used on ships, for instance, as adhesives, in acoustic tile, nonskid surfaces, etc.

Although epoxy resins are flammable (Conley and Quinn, 1975; Lyons, 1970), their flammability can be reduced considerably by the use of a variety of phosphorus or halogen-containing additives or reactive monomers. The need for fire retardance in epoxy resins has been relatively small to date, and the amount of fire retardant epoxy resins sold yearly is small. Because there is relatively little use of these resins a major fire hazard has not resulted.

4.4.3 Specialty Plastics

This somewhat arbitrary subdivision comprises those materials that are relatively high priced, have certain particularly outstanding properties, and are produced in relatively small volumes for specialty applications. Until the prices of these polymers are reduced, there is little prospect of their large-scale usage in ships. The materials in this group fall into two general categories: (1) the aromatic and hetero-aromatic polymers generally used for their resistance to temperature, chemicals, and/or combustion (there is a detailed discussion in Volume 1 of such polymers); (2) the polyimides which are discussed below.

4.4.3.1 Aromatic Polyimides

Aromatic polyimides are expensive, but should find increased use as flame resistant insulating materials. They are synthesized from an aromatic dianhydride and an aromatic primary diamine. They are used primarily for moldings, composite matrix resins, films, and foams. The films are the best non-halogenated fire-resistant films available. They are clear, amber colored, fairly flexible, and have high tensile strengths. Fibers have been spun from polyimides and their mechanical properties have been investigated exhaustively. A typical polyimide repeat unit is poly(oxydiphenyl pyromellitimide) (Kapton®).

Various woven forms of at least one polyimide fiber have been characterized thoroughly. Kevlar, a fiber spun from the above polymer, demonstrated excellent thermal aging characteristics (Opt and Ross, 1969; Ross, 1967).

4.5 Composites and Laminates

4.5.1 Introduction

An enormous quantity of composites and laminates, made from various synthetic plastics, wood products, various fillers, reinforcing agents, and adhesives, is produced annually in the United States. Like fiber blends, composites and laminates may pose special fire hazards. In many cases, the exact composition of these materials is proprietary. In any case, however, their fire performance cannot be predicted from a knowledge of the performance of the components. The unsaturated polyesters, phenolics, epoxies, and amino resins are used most often in the manufacture of composites and laminates. High-temperature resins such as polyimides are receiving increased attention in composite structures. Chapter 5, Volume 6, Aircraft: Civil and Military of this series discusses briefly some of the more advanced materials under investigation for use in composites.

4.6 Foams

4.6.1 Introduction

Foamed polymers are extensively used for seat cushions, padding, insulation, etc. Foamed polymers pose special fire hazards and their fire safety aspects require special consideration. Polymeric foams are generaly complex multicomponent systems which may, in addition, contain fibers and various fillers. Polymeric foams may be rigid or flexible.

In the first of two foam types, the cellular structure is obtained with the aid of a blowing or foaming agent. This may be either a liquid which vaporizes during the polymerization process, or a solid which decomposes to give off gas. Some gas (carbon dioxide or water vapor) may also be formed as part of the polymerization reaction.

Syntactic foams are the second type. These are essentially polymers which contain tiny hollow spheres of another polymer or glass as filler.

Flexible foams generally contain an open cell structure. Rigid foams are usually closed cell foams.

Since the burning of a polymeric composition can only occur on the surface, the surface area available for combustion is important in determining the rate of combustion and, therefore, the intensity of the flame. Thus, a film burns more easily than a thick molded part, and a foam burns more readily than a solid polymer of the same composition. The burning of polymer foams, therefore, differs in several respects from the burning of solid polymers.

Since a greater surface is exposed to air, the rate of pyrolysis and burning of a foam is greater than that of the base polymer. The low thermal conductivity of foams tends to concentrate the heat on the surface of the structure, rather than dissipating it to underlying material or substrate. The result is a rapid heatup and pyrolysis of the surface material when exposed to a flame. This often leads to an extremely rapid flame spread.

Other factors may moderate the surface effect considerably; for example, the small amount of potentially flammable material per unit volume in low-density foams results in a very small amount of total heat to cause flame propagation. Thus, if the foam is made from a thermoplastic resin such as polystyrene, the heat of a flame rapidly melts the foam adjacent to it. The material then may recede so fast from the flame front that there is no real ignition. However, when polystyrene foam is caused to burn, it does so quite rapidly and evolves much smoke. These and other factors have led to considerable concern regarding the use of foams in many applications where relatively large areas and/or volumes of materials are involved.

A highly crosslinked thermoset foam, on the other hand, behaves in an entirely different manner. Since little or no melting occurs, the surface does not recede from the flame front and the foam is rapdily ignited. A highly insulating char is formed, protecting the remainder of the material from the flame. Thus, a relatively flammable solid is converted into less flammable carbon. Since carbon itself is combustible, the continued raidant heat can generate continued combustion, but re-radiated energy from the char and the low density of the surface char generally inhibit sustained burning.

4.6.2 Rigid Foams

Some limited applications for rigid foams are insulation, rigid components, and sound barriers. Because of their limited application in ships, some of the more common rigid foams such as polystyrene will not be discussed here.

Rigid foams may be fire retarded in several ways. These may include the use of fire suppressant additives such as organic phosphate derivatives (Bagnoli, 1973; Ishizuka et al, 1973; Batorewicz, 1973), either in the foam formulation or applied to the formed surface in the form of coatings. Both techniques effectively reduce the burning rate of the foam surface. A more recent practice is the introduction of an inorganic filler (i.e., alumina trihydrate), to control flame spread.

An alternate method of achieving fire retardation in foams consists in making them of fire retardant polymers. In one approach, halogens such as chlorine or bromine are introduced into the backbone of the basic polymer (Shone, 1973; Stastny et al, 1973, Yamaguchi et al., 1973, Anon, 1973). Another approach consists of using high-char-yield materials such as polyisocyanurates, polyimides, polybenzimidazoles, polyquinoxalines, and polyphenylenes. These polymers directly yield fire retardant foams capable of passing present-day flammability tests without the need of additional fire suppressant additives.

The application of cement, gypsum board, plaster, or other inorganic surface coatings onto the surface of a rigid foam provides another method of controlling their potential fire spread and ease of ignition.

Rigid polyurethane foam and expanded polystyrene are the most widely used materials in the thermosetting and thermoplastic class, respectively.

Among natural foams, cork is the only important material and is quite hazardous.

4.6.2.1 Rigid Polyurethane Foams

The market for rigid foams in ships is dominated by polyurethane. They are used predominantly as insulation in cargo holds and as sound barriers. In the cargo area of tankers and on naval vessels, these foams are used as tank and pipe insulation. On naval vessels, rigid foams are used as thermal insulation for "reefer" spaces, but only if encased in metal sheet.

The cellular nature and low thermal decomposition temperature of polyurethane foams generally influence their flammability. The high heat flux generated by surface combustion, coupled with low thermal conductivity and high flame temperatures, can cause almost instantaneous conversion to flammable gases. These processes in turn often result in extremely rapid surface flame spread.

In general, some degree of fire retardance is imparted to polyurethane foams by the chemical incorporation of halogens and/or phosphorus compounds into the material. The use of phosphorus compounds in fire retardant polyurethane foams leads to high char formation combined with easy processing because of the relatively low viscosity of most phosphorus compounds. This combination of desirable properties has made phosphorus compounds, with or without halogen, the most widely used fire retardants in polyurethane technology. Reactive phosphorus compounds, such as Fyrol 6 (Stauffer Chemical Company) are used extensively. They are added directly to the polyol.

$$[CH_3CH_2O]_2 \overset{\overset{\textstyle O}{\|}}{P}-CH_2-N [CH_2CH_2OH]_2$$

Polyurethane foams may be fire retarded also by the incorporation of non-reactive additives which act as fillers or plasticizers. Non-reactive additives have not been used extensively in polyurethane technology because of their tendency to migrate from the foam under many conditions of extended use.

Although polyurethanes themselves are nontoxic, the pyrolytic combustion gases have been shown to contain considerable quantities of toxic gases. Significant amounts of hydrogen cyanide have been detected in polyurethane combustion products (Sumi and Tsuchiya, 1973), although the relative toxicity hydrogen cyanide in gaseous mixtures that contain large amounts of carbon monoxide has not been definitely established.

4.6.3 Flexible Foams

Flexible foams can be made from practically any elastomer. They are used in a variety of applications, the most important being seat cushioning, carpet underlay, and carpet backsizing. Flexible foams, predominantly polyurethanes, are used extensively as cushioning materials.

When a chemical blowing agent is used in an elastomer recipe, the foam rubber is generally referred to as sponge rubber. Sponge rubber is made mostly from styrene-butadiene rubber, although silicone and fluorocarbon (Viton) sponge rubbers are also available.

Latex foam rubber is made by beating air into compounded rubber latex. Fluorocarbons are used as (additional) foaming agents in some processes. A gelling agent such as sodium silicofluoride or ammonium acetate may be used (Morton, 1973). Natural or styrene-butadiene rubber, or blends of the two, are widely used.

Flexible polyurethane foam constitutes a high percentage of all elastomeric foams. According to Gmitter and Maxey (1969), flexible slab polyurethane foam accounts for about two-thirds of all flexible foam. The method of preparation of flexible polyurethane is not essentially different from that of the rigid foam. Flexibility is achieved by appropriately varying the molecular weight and the functionality of the polyols used in preparation. Most currently available flexible polyurethane foams are made by the reaction of an isocyanate with polyethers.

Methods of "fire retardation" of flexible polyurethane foams are essentially the same as those used with rigid polyurethane foams. Unfortuantely, flexible polyurethane foams burn readily even when fire retarded. A totally satisfactory solution to the pressing problem of how to make a fire retardant flexible foam for cushioning has not yet been developed. (See Volume 7 for a discussion of the Vonar® system).

4.7 Wood and Wood Products

4.7.1 Introduction

Wood and wood-based products have historically been among the most widely used materials for manufacturing ships of many types. It is reported (Genesis 6:14—16) that Noah was comissioned to use gopher wood in the construction of the Ark. Replacement of wood structures by steel and aluminum has been prevalent for many years in large and military vessels. Recent pleasure or small craft have not used wood very frequently, since polyester/glass combinations have proved to be more easily fabricated. Considerable technical information is available on methods for the treatment of wood to reduce the flammability and initial rate of heat release, and to make it self-extinguishing. However, these treatments have not been widely applied in ships because of the limited use of wood in new construction. (The fire-retardant treatment of wood is discussed in detail in Volume 1 of this series).

4.8 Fire Retardant Coatings

4.8.1 Introduction

The use of fire retardant coatings is one of the oldest methods for protecting flammable and non-flammable substrates from reaching ignition or softening temperatures. The two main types of fire retardant coatings are the intumescent and non-intumescent varieties. Fire retardant coatings can be used to reduce the flame spread characteristics of almost any type of organic substrate. Although fire retardant coatings are currently little used in transportation, the present public pressure for reduced flammability and smoke-generating ability could lead to increased use of such coatings to decorate the interior of ships. Because of this possibility, a short review of fire retardant coating technology is included.

4.8.2 Paints and Coatings

Non-intumescent coatings do not provide the same degree of fire protection to the substrate as do intumescent coatings; nevertheless, they do not enhance the spread of flame by rapid combustion, or contribute a significant amount of fuel to the fire.

4.8.2.1 Alkyd Coatings

The most frequently used fire retardant coatings are based on chlorinated alkyd resins prepared predominantly from chlorendic anhydride or tetra-chlorophthalic anhydride. By using the proper chlorinated acids, coatings can be made which have properties comparable to those of conventional coatings in addition to being fire retardant. It is this high performance and relatively low cost that has made chlorinated alkyd coatings so successful.

The addition of fire retardant additives to alkyd resins is also commonly employed. Halogenated additives such as chlorinated paraffins are the most commonly used additives in these coatings because of their low cost. Antimony oxide is the most commonly used synergist in these applications.

4.8.2.2 Miscellaneous Coatings

Other polymers have been used much less in coatings. Such coatings are based upon urethanes (Saunders, 1962) and epoxy resins (Lyons, 1970 and Anon, 1971). Additionally, fire retardant heat cured coatings based upon melamine/formaldehyde and phenol/formaldehyde resins may find significant use as a char resistant coating on factory coated wood.

4.8.3 Intumescent Coatings

Intumescence is defined as "an enlarging, swelling, or bubbling up (as under the action of heat)" (Webster's Third New International Dictionary, 1961). Insulating intumescent coatings are used to protect vulnerable substrates from reaching ignition temperature. They also protect non-flammable substrates, such as metals, by preventing them from reaching softening or melting temperatures. A thorough review of intumescent coatings was published by Vandersall (1971).

Conventional intumescent coatings contain several key ingredients which are necessary to bring about the intumescent action. An intumescent catalyst is used to trigger the first of several chemical reactions which occur in the coating film. A carbonific compound is included; it reacts with the intumescent catalyst to form a carbon residue. A spumific compound is included; it decomposes, producing large quantities of gas which cause the carbonaceous char to foam into a protective layer. A resin binder forms a skin over the foam and keeps the trapped gases from escaping. Apart from these key ingredients, intumescent coatings may also include other ingredients used in conventional coatings such as pigments, driers, leveling agents, and thinners.

Unfortunately, many intumescent compositions have one or more drawbacks such as: poor aging, poor weathering, poor humidity resistance, poor flexibility, and high cost.

Non-conventional intumescent coatings are those in which the elements of intumescence are built into the resin. A few such coatings have been recently described; for example, a clear intumescent epoxy coating (Blair et al., 1972) has been prepared by the reaction of triphenyl phosphite with an epoxy resin prepared from epichlorohydrin and bisphenol-A. The coating was prepared by adding the amine catalyst to the premixed epoxy-(triphenyl phosphite) resin just before it was applied. The coating, as applied, consisted of 100 percent solids, as found in the final product.

A similar fire retardant clear intumescent urethane coating was reported by Clark, et al. (1967, 1968). A moisture-curing polyurethane was prepared from pentachlorophenoxy-glycerol ether, triethylene glycol, and toluene diisocyanate in a solvent.

4.9 Conclusions and Recommendations

Conclusion: Given sufficient oxygen and thermal energy input, all organic polymers will burn. *Recommendation:* Increase the development effort on char-forming systems with particular emphasis on lowering the fabrication cost.

Conclusion: Many synthetic organic polymers burn in a manner different than that of the more familiar natural polymers such as wood, paper, cotton, or wool. *Recommendation:* Initiate programs to increase basic knowledge of the relationship between the chemical and physical properties of polymers, the fire dynamics parameters, and the way these relations are affected by aging.

Conclusion: Some uses of polymeric materials in ships can seriously augment the fire hazard. *Recommendation:* Concern exists about potential fire hazards associated with the rapidly increasing use of polymeric insulating foams in ship construction. Support approaches to improve the fire safety of the high-volume low-cost polymers.

Conclusion: Polyvinyl chloride and polyolefins have been and will remain in the near future the primary insulating materials used in wire and cable insulation. Many factors, but primarily economics, dictate their continued use. *Recommendation:* Define problems, if any, associated with new PVC wire insulation formulations. Develop approaches to surmount these difficulties. Develop comparative data on "fire-resistant improved" PVC formulation(s).

Conclusion: Fluoroolefins will receive increased attention. They are limited to particular applications because of the high materials cost. There may be certain problems caused by the toxicity of combustion products associated with their use. *Recommendation:* Define and attack any problems associated with fluorine-containing polymers.

Conclusion: Polyphosphazenes are the only new more fire retardant materials currently approaching commercialization and are emerging as a new class of higher priced materials to consider for use in fire retardant applications. *Recommendation:* Encourage future development of polyphosphazenses, defining advantages and problems.

Conclusion: Silicones and polyimides (primarily in film form) are two classes of commercial cable-insulating materials which do not depend on halogens (either as an addition or as part of the structure of the polymer) to provide fire-resistant properties. However, they do not fit into the same economic framework as PVC and polyethylene. *Recommendation:* Develop novel non-halogen approaches to the flame retarding of polymers.

Conclusion: The fire flammability safety of many polymers has been improved by the incorporation of hydrated alumina and/or compounds containing halogens, phosphorus, and/or antimony. *Recommendation:* Create an overall program which will categorize and communicate the goals and results of government-supported work on the fire safety of polymeric materials, particularly addressing the use of additives.

Conclusion: The polymeric materials with improved flammability characteristics which are available today may have deficiencies such as high costs, fabrication difficulties, and formation of toxic and corrosive combustion products. *Recommendation:* Initiate programs to determine the relationship of chemical and physical components of polymeric materials to the evolution of smoke and toxic gas formation.

Conclusion: The use of intumescent coatings can enhance the fire safety of some polymeric products, but may have limited usefulness in a marine environment because of adverse effects of moisture and salt water. *Recommendation:* Expand materials and application studies on intumescent coatings, with emphasis on lowering the cost and improving coating performance in an adverse environment.

REFERENCES

Al'Shits, I.M., Flis, I. E., J. Appl. Chem. (USSR) *34*, 618 (1961).

Anon., Plastics and Rubber Weekly, No. 508, p. 11, Nov. 30 (1973).

Bagnoli, C. D., Spanish Patent 379, 750; Chem. Abs. 78 59327a (1973).

Batement, R., ed., "The Chemistry and Physics of Rubber-like Substances," John Wiley, New York (1963).

Batorewicz, W., Germany Patent 2,237,591; Chem. Abs. 79, 32459d (1973)

Benisek, L., International Dyer and Textile Printer, 414—19, April 7, (1972).

Benisek, L., Melliand Textilever., 8 (6), 931—34 (1972a).

Benisek, L., Textileveredlung, 8 (6), 318—326 (1973).

Billmeyer, F. W., Jr., "Textbook of Polymer Science," 2nd ed., Wiley Interscience, New York (1971).

Blair, N. D., Witschard, G. and Hindersinn, R. R., J. Paint Tech., 44, 75—82 (1972).

Brydson, J. A., "Plastics Materials," 3rd ed., Van Nostrand, New York (1975).

Clark, C. C., and Krawczyk, A., U.S. Pat. 3,365,420 (1968).

Clark, C. C., Krawczyk, A., Reid,G. C., and Lind, E. V., Paint and Varnish Prod., 56—59, (April 1967).

Conley, R. T., and Gaudinana, R. V., "Thermal and Oxidative Degradation of Polyamides, Polyesters, Polyethers, and Related Polymers" in Thermal Stability of Polymers, Vol. I, p. 347, ed. R. T. Conley, Marcel Dekker, N.Y. (1970).

Conley, R. T., and Malloy, R., "Vinyl and Vinylidene Polymers" in: Thermal Stability of Polymers, Vol. I, p. 223, ed. R. T. Conley, Marcel Dekker, N.Y. (1970).

Conley, R. T., and Quinn, D. F., "Retardation of Combustion of Phenolic, Urea-Formaldehyde, Epoxy, and Related Resins Systems" in: Lewin, Atlas, and Pearce (1975).

Dalzell, D. A., and Nulph, R. J., Society of Plastics Engineers, 28th Annual Technical Conf., Vol. XVI, p. 215, New York (May 1970).

Drake, G. L., Jr., "Fire Resistant Textiles," in: Kirk-Othmer Encyclopedia of Chemical Technology, 2nd ed., Wiley-Interscience, (a) Vol. 9, pp. 300—15 (1966); (b) Suppl. Vol., pp. 944—64 (1971).

Einhorn, I. N., "Fire Retardance of Polymeric Materials," J. Macromol. Sci. Revs. Polymer Technol. D1(2), 113—184 (1971).

Essenhigh, R. H., and Howard, J. B., Ind. and Eng. Chem. 58, 15 (1956).

Fabris, H. J., and Sommer, J. G., "Flame Retardation of Natural and Synthetic Rubbers," in: Flame Retardancy of Polymeric Materials, Kuryla and Papa (eds.) Marcel Dekker (1973).

Floyd, K. L., and Taylor, H. M., "Industrial Applications of Textiles," *Textile Progress, A Critical Appreciation of Recent Developments,* Vol. 6, No. 2, The Textile Institute, Manchester, 1974.

Foy, G. F., "Engineering Plastics and Their Commercial Developments," Advances in Chemistry, American Chemical Society Monographs, No. 96 (1969).

Freidman, M., et. al., Text. Res. J., 43, 212—17 (1973).

Fristom, R. M., Chem. and Eng. News, 150, Oct. (1963).

Gmitter, G. T., Fabris, H. J., and Maxey, E. M., in: Flexible Polyurethane Foam, Chapter 3, Vol. I, Part I, pp. 109—226, Frisch and Saunders (eds.) (1973).

Gouinlock, E. V., Proter, J. F., and Hindersinn, R., J. Fire and Flamm. 2, 206 (1971).

Gross, D., Loftus, J. J., Lee, T. G., and Gray, V. E., "Smoke and Gases Produced by Burning Aircraft Interior Materials," Building Science, Series 18, National Bureau of Standards (1969).

Hagnauer, G. L., and Schneider, N. S., J. Polymer Sci., A-2, 10, 699 (1972).

Hecker, K. C., Rubber World, 159 (3), 59 (1968).

Hilado, C. J., "Flammability Handbook for Plastics," Technomic Publishing Company, Stamford, Conn. (1969).

Hindersinn, R. R., and Wagner, G. M., "Fire Retardancy" in: EPST 7, 1 (1967).

Hooker Chemical Corp., (a) Preliminary Data Sheet No. 349, "Dechlorane 604" (August 1970); (b) Techn. Bull. Dechlorane Plus, 602 and 604, Fire Retardant Additives for Elastomers (1970).

Howarth, J. T., Lindstrom, R. S., Sheth, S. G., and Sidman, K. R., "Flame Retardant Additives," Plastic World, p. 64—74 (March 1973).

Ishizuka, Y., et al., Japanese Patent 72,39,197, Chem. Abst. 78, 98477u (1973).

Kaswell, E. R., *Wellington Sears Handbook of Industrial Textiles,* p. 32, 65, Wellington Sears Co., Inc., New York (1963).

Kennedy, J. P., and Tornquist, E. G. M., eds., "Polymer Chemistry of Synthetic Elastomers," Interscience, New York (1968).

Krekeler, K. and Klimke, P. M., Kunstoffe 55 (10), 758 (1965).

Laur, T. L., and Guy, L. G., Rubber Age, 102 (12), 63 (1970).

Lyons, J. W., "The Chemistry and Uses of Fire Retardants," Wiley-Interscience, New York, (1970).

Madorsky, S. L., Thermal Decomposition of Organic Polymers, Interscience Publishers, pg. 160 (1964).

Mark, V., U.S. Patent 3,940,366 to General Electric (1975).

Morton, M., ed., "Rubber Technology," Reinhold, New York (1973).

Nametz, R. C., Ind. Eng. Chem., 623, 41—53 (1970).

Nouvertne, W., U.S. Patent 3,775,367 to (Farbenfabriken Bayer) (1975).

Opt, P. C. and Ross, J. H., Appl. Polymer Symp., No. 9, 23—47 (1969).

Pepe, A. E., German Patent 1,936,345 (1970).

Polsar Inc., Data Sheet/600-R "Polysar SS 1905" (1970).

Roberts, A. H., Haig, D. H., Rathsack, R. J., J. Appl. Polymer Sci. 8, 363 (1964).

Ross, J. H. and Opt, P. C., Textile Res. J., *37,* 1050 (1967).

Ross, J. H., Safe Engineering, 2 (4), 16 (1973).

Rosser, W. A., Wise, H. and Miller, J., *Seventh International Symposium on Combustion,* Butterworth Science Publications, London, p. 175 (1959).

Saunders, J. H., and Frisch, K. C., "Polyurethanes," Part I, Chemistry, Interscience, New York (1962); Part II, Technology (1964).

Shone, R., et al., U.S. Patent 3,704,256, Chem. Abst. 78, 30775j (1973).

Stastny, F., et al., German Patent 2,134,211, Chem. Abst. 78, 98, 642y (1973).

Stern, H. J., "Rubber: Natural and Synthetic," Palmerton, N.Y., (1967).

Sumi, K., and Tsuchiya, Y., J. Fire and Flammability, 4, 15 (1973).

Sunshine, N. B., "Flame Retardance of Phenolic Resins and Urea- and Melamine-Formaldehyde Resins," Flame Retardancy of Polymeric Materials, Kuryla and Papa (eds.), Marcel Dekker (1973).

Tesoro, G. C., and Mesier, C. H., Jr., Textile Res. J., 40, 430–36 (1970).

Texas-U.S. Chemical Co., Belgian Patent 729,226 (1964); British Patent 1,010,145 (1964).

Trexler, H. E., "The Formulation of Nonburning Elastomer Compounds," Rubber Chem. Tech. 46, 1114–1125 (1973).

Vandersall, H. L., "Intumescent Coating Systems, Their Development and Chemistry," J. Fire and Flammability, 2, 97–140 (1971).

Weiner, S. A., American Chemical Society, Div. of Organic Coatings and Plastics Chemistry Preprints 34 (1), 55 (1974).

Whitby, G. S., Davis, C. C., and Dunbrook, R. F., eds., "Synthetic Rubbers," John Wiley, New York (1954).

Winspear, G. G., ed., "The Vanderbilt Rubber Handbook," R. T. Vanderbilt Company, New York (1972).

Yamaguchi, K. et al., Japanese Patent *73,* 11,341, Chem. Abst. 78, 160, 507j (1973).

CHAPTER 5

SPECIFICATIONS, STANDARDS AND TEST METHODS

5.1 Introduction

A complex spectrum of test methods, standards, specifications, codes, and related regulations govern the role of materials in marine fire safety. The complexity of this system and the large number of components involved present serious difficulties for the producers of polymeric materials and products and for the users. Ships have unique problems associated with fires, but continued improvement of specifications and standards and their implementation in regulations and contracts over the years has resulted in decreased fire losses. However, introduction of new polymers and increased substitution of polymeric materials for metals requires that the specifications and standards be reevaluated.

The objective of all fire safety activities is to reduce loss from fires. The precise definition of loss is itself a difficult and controversial problem, but the focus here will be on two major elements, human death and injury, and the destruction of material property.

Both the Coast Guard and the Navy have established test methods, standards, and specifications which have their origins in the following activities:

- the statutory efforts of government regulatory agencies to reduce the frequency and severity of fires through mandatory standards;
- the voluntary efforts of national and international associations concerned with the development of fire test methods, recommended safety practices, and dissemination of authoritative fire safety information.

A test method is a procedure for measuring a property or behavioral characteristic of a material, a product, or assembly as an aid to predicting its performance in an intended application. Specifications find their principal application where they are used to define the properties of a material, product, or structure being procured. They establish the level of performance that must be met and the test method(s) by which such performance is to be measured. When incorporated into a purchase contract, the specification becomes part of a legal document, enforceable in the courts. Ideally, a fire test method should be a procedure that can be used to predict the performance of a material, product, structure, or system under fire exposure conditions which can reasonably be anticipated in the intended application.

Test methods may measure different elements of *fire hazard* such as:

1. Ease of Ignition
2. Surface flame spread
3. Heat release
4. Smoke evolution
5. Toxic gas formation
6. Fire endurance
7. Thermal transmission

Acceptance tests are generally classified according to the intended end use of the product: this frequently requires compliance with specific regulations. Finally, tests are frequently classified according to size, and designated by such descriptive and nonquantitative terms as large-scale, full-scale, sub-scale, small-scale, and laboratory-scale. From a fundamental standpoint, this is the least useful type of classification, but from a practical standpoint, the size of the test may be critically important.

A principal measure of the fire hazard in a particular type of ship is given by the results of a series of tests that have been developed to evaluate the choice of materials to be used in that system. With these test results, design engineers may compare their choices of materials to the specification limits of the standards. The specification limits may be chosen to limit the choice to the few best materials, or only to eliminate the worst materials in a given category.

One of the most common criticisms to which any laboratory test method may be subjected is that it does not fully simulate an actual, real-life situation. If a material were to be exposed to fire risk in only one precisely defined set of circumstances (size, orientation, type of igntion source, method of applying the ignition source, ventilation conditions, environmental conditions, etc.), it would be obvious how to test for fire hazard. A series of candidate materials would be evaluated under this set of circumstances and materials found to be satisfactory or unsatisfactory could be noted. One then could evaluate smaller samples of the same series of materials by means of another proposed smaller-scale test method, and the smaller-scale test method could be validated for future use if the results of the two procedures correlated.

Unfortunately, this idealized approach is frequently not directly applicable. The principal reason is that a material of interest may be exposed to fire risk in a wide variety of ways rather than by a single, well-specified set of circumstances. Thus, an unacceptably large number of experimental fires would be needed to explore fully all the permutations and combinations of the variables. A second reason is that in some cases even a few realistic fire tests would not be practical, if, for instance, each experiment required destruction of a ship or even a small boat. Nor would a best be permissible if the experiment required exposure of human beings to lethal fumes.

This chapter aims to enumerate and evaluate the current fire hazard test methods, specifications, and standards used in the selection of materials for ships. An additional objective is the correlation of these test methods with the other facets of fire safety offered in other chapters of this volume. A listing of the regulations and specifications that have been adopted for the use of the different types of ships under Coast Guard and Navy jurisdiction will be found in this chapter.

For a more thorough discussion of test methods, specifications, and standards, the reader is referred to Volume 2, Test Methods, Specifications and Standards of this report.

5.2 Organizations Involved in Regulations and Requirements

5.2.1 Coast Guard and Nonmilitary Organizations

Under the overall cognizance of the Coast Guard, within the commercial maritime community, four primary bodies are concerned with a vessel's fire safety specifications. These are:

> Owner.
> Classification Societies.
> National Administration (Country of Registry).
> Intergovernmental Maritime Consultative Organization.
> (IMCO).

Each of these bodies represents an interest in the safe design and operation of the vessel.

IMCO is a specialized agency of the United Nations. It serves as the depository for the 1960 Safety of Life at Sea Convention (SOLAS), which is a building and exits code for ships. The United States, among other maritime countries, is signatory to that treaty and all utilize its provisions as a minimum basis for their national regulations. Signatory nations develop regulatory frameworks either adopting by reference or improving upon the basic provisions found in SOLAS 1960.

The classification societies, which assist in setting insurance rates for the underwriters by setting forth and inspecting to certain standards, also have a strong interest in the setting of standards.

Ship owners often exercise the prerogative to upgrade or demand a higher standard than that required by convention or by the federal government.

5.2.2 Navy

The Navy formulates its own regulations, which are developed at the branch level and then formalized at the higher levels.

5.3 Specifications and Test Methods

5.3.1 Coast Guard

5.3.1.1 Ships (Over 60' in Length)

5.3.1.1.1 Hull Materials, Title 46, Subpart CFR 177.10-5

Various provisions of Title 46 "Code of Federal Regulations, Shipping," require the hull, decks, and deckhouses of merchant vessels to be constructed of steel or equivalent material. The equivalence clause permits, in some instances, the utilization of aluminum. Certain types and sizes of vessels which traditionally have been constructed of wood are permitted to be constructed of glass-reinforced plastic (GRP). The latter has been extensively used in small passenger vessels which carry more than 6 but less than 150 passengers. In 1972, it was recognized that GRP hulls constructed of general purpose resins presented a potentially serious fire hazard. Wood has been the preferred material of construction for this type of vessel in the past; therefore its fire hazard properties were used as benchmarks by which future requirements of GRP hulls could be developed. Specific parameters of importance were ease of ignition, flame spread, and heat of combustion.

Title 46 CFR 177.10-5 requires that the resins used comply with MIL-R-21607 after a 1-year exposure to weather, if GRP is used as a primary structural element in the vessel. In this test procedure, the GRP sample is fabricated in sheets, with 40 sheets of resin-impregnated glass cloth press-cured to a thickness of 12.7 mm. Samples of this composite are tested according to Federal Standard 406, Method 2023 (1961). In this test, a 5″ X ½″ X ½″ sample, mounted vertically, is surrounded by a heating coil. Two spark plugs ignite any combustible gases arising from the heated specimen. The burn time and flame travel are reported. For Grade 1 materials, minimum average ignition time is 55 seconds. For Grade 2, minimum average ignition time is 70 seconds. The corresponding maximum for average burning times are 125 and 65 seconds respectively.

This test procedure is expected to be abandoned soon in favor of a new test procedure now under investigation by the National Bureau of Standards. (See 5.3.1.1.5).

5.3.1.1.2 Deck Coverings (46 CFR 164.006)

This test requires that a sample whose organic carbon content may not exceed 0.12 gm/cm^3, at least 6 inches square and ¼ inch thick, placed on an iron plate, be exposed in a furnace that is controlled according to the standard fire exposure curve, reaching 1700°F at the end of 1 hour (per ASTM E-119). The temperature of the unexposed side is noted at 5-minute intervals and may not rise higher than 250°F above the original temperature at the end of 15, 30, and 60 minutes. Excessive cracking, buckling, or disintegration may be considered cause for rejection.

The test for smoke requires that each sample be placed on an iron plate in a furnace whose temperature is limited to the standard decking curve, reaching 1325°F at the end of 1 hour. Average light transmission may not be less than 90 percent at 15 minutes, 60 percent at 30 minutes, and 50 percent at 60 minutes.

5.3.1.1.3 Bulkheads (46 CFR 164.008)

Bulkheads require a test for noncombustibility, (essentially a modified ASTM E-136), which disqualifies most materials having an organic material content in excess of 6—8 percent by weight. In addition, bulkheads are required to undergo a modified ASTM E-119 test to determine thermal penetration. The maximum requirement for a bulkhead is a classification that is known as A-60. The "A" designates that it has to be steel and capable of remaining in place for 1 hour, and the 60 designates a period of 1 hour, during which the temperature of the unexposed side cannot exceed 250°F (139°C). "B" class bulkheads are required to exhibit structural integrity for 30 minutes, prevent the passage of flame, and have a thermal rating of 0 to 15 min. "C" class bulkheads are only required to be noncombustible by test. (ASTM E-136).

5.3.1.1.4 Vessel Furnishing and Decorative Elements

While the bulkheads and certain other structural components of ships are required to be noncombustible, interior finish, furniture and deck covering are permitted to be combustible to some extent. Some of the test methods presented in this section are currently being reviewed for replacement. A discussion of the proposed new test methods is presented later in this chapter.

a. Interior Finish (46 CFR 164.012)

The maximum allowable thickness of any combustible interior finish is 0.075 inch. The applicable test method is ASTM E-84. Interior finishes must have a flame spread of 20 or less, and smoke generation no greater than 10 when bonded to a ¼ inch asbestos cement board. The areas common to all vessels that must comply with this structural fire protection requirement are the corridors, stairways and stairtowers, and hidden or concealed spaces.

b. Furniture

Furniture is divided into two primary categories — fire-resistant and non-fire-resistant. The so-called fire-resistant furnishing is a misnomer; although the frames are required to be metal, the padding and upholstery only have to be self-extinguishing when tested according to ASTM D-1692. (ASTM D-1692 will be removed from ASTM Book of Standards after the 1978 edition).

It should be noted that fire-resistant furnishings are required only in certain areas aboard passenger vessels.

Standard ASTM D-1692 was developed for the testing of cellular plastics, and specifies that the material, mounted horizontally on a supporting screen, be heated at one end for 60 seconds by flame from a fish-tail bunsen burner. A sample is considered to be "self-extinguishing" if any burning does not go beyond a mark 125 mm. from the point of ignition. This test is not considered by fire experts to be acceptable to forecast the performance of materials in real fire situations. The horizontal positioning of the sample does not take into account the performance of materials that are vertical and thus burn at a much faster rate, and possibly in a different manner. Other test methods, such as ASTM E-162, are considered more representative of a real fire situation.

c. Carpets

International convention requires the carpets to be wool or equivalent. The equivalency is determined by the ASTM E-84 test method, and requires a flame spread value of 75 or less to pass.

This test procedure is no longer considered a valid method for testing carpeting. The new NBS Flooring Radiant Panel Test (NFPA Standard No. 253-78) is now gaining acceptance as a preferred test method.

d. Thermal Insulation

In accommodation and service spaces, low-density organic foam is permitted to be used only as insulation surrounding refrigerated compartments, provided it is encapsulated in steel. In cargo areas, however, these foams are often used in refrigerated space. In many instances, it is foamed in place and has no protective outer cover, thereby posing a potential ignition and high heat-release source.

Polyurethane foam is currently used as cryogenic insulation for Liquefied Natural Gas (LNG) tanks. The foam must be self-extinguishing in accordance with ASTM D-1692, (which is no longer a Standard), or the hold containing the foam must be inerted. (See comments in Section 5.3.1.1.4 b,).

e. Electrical Insulation

Electrical insulation fires on shipboard are viewed with grave concern by both the Coast Guard and Navy because of the serious losses that have occurred by propagation along the cableways. Wire and cable traverse the ship from end to end and bottom to top and thus fires must be stopped at the decks and bulkheads by paying special attention to the penetrations in decks and bulkheads.

The principal shipboard cable in use today is insulated with polyvinyl chloride, which is protected by wire braid. The Coast Guard now requires that cable meet the fire test provisions of IEEE Standard 45, with reference to IEEE Standard 383. (Refer to Title 46-Shipping, Subchapter J. 46 CFR 111.60). This method requires that a ribbon gas burner with a heat equivalent output of 70,000 BTU per hour impinge on the vertically-oriented cables. There is consideration now being given to higher heat fluxes.

A program is currently underway at the David W. Taylor Naval Ship Research and Development Center for the development of highly fire-retardant shipboard electrical cables. Smoke and toxic gas emission from burning electrical insulation are also major problem areas being addressed in the program.

Other projects being conducted by the Navy under the general program, "Cable and Wireway Fire Protection" are:

- Fire Stop Materials (Bulkhead penetration)
- Electrical Cable and "Ampacity" Tests (Cable current rating)
- Cable Coating and Wrapping Materials (Total run fire protection)
- Improved Cable Materials (New Insulation Materials)

The results of these investigations could lead to new test methods for shipboard wire and cable, as well as new methodology for the design of layouts and penetrations.

f. General

For many years, Federal Test Method Standard No. 406 methods have been specified by both the Coast Guard and the Navy. The last issue of this standard was published in 1961. The Navy recently asked a committee of fire experts to revise this standard and replace it with ASTM standards. It can therefore be anticipated that new standards will be promulgated to specify the flammability and smoke emission of combustible ship materials.

5.3.1.1.5 Proposed New Test Method

A new flame test procedure, labelled ISO/TC 92 N 453 or ISO/TC 92/WG 4 N 243, has been proposed by the International Organization for Standardization (ISO). It is now being investigated by the National Bureau of Standards at the request of the Coast Guard. The proposed test method uses a radiant panel as a source of heat at a temperature of 750°C. Specimen samples, 800 $\times$ 150 mm, are exposed in either a floor, wall, or ceiling position. A pilot flame is also used. Time of ignition is noted, as well as the time that the flame front passes marks at 100 mm, 150 mm, 200 mm, etc., until the flaming ceases or the flame reaches the end of the specimen. The potential advantage of this system is that materials can be tested in the orientation in which they are intended to be used. One possible fault is that the temperature above the burning samples is not monitored as in ASTM E-162. Thus, a material which chars and does not present a flame front may emit highly combustible gases which would not be detected. This would allow a potential flash-over condition.

5.3.1.2 Small Boats

A perusal of the Coast Guard specifications for small boats fails to reveal any specifications for flammability of materials. Although fatalities, injuries, and mone- ary losses are small relative to land transport and building, construction increasing amounts of plastics are being used in the construction and outfitting of small craft thus increasing the fire hazard. Further, pleasure craft are particularly susceptible to fires because of the generally inexperienced crews. Thus, careful monitoring of material uses and fire statistics are very important if fire safety problems are to be rationally studied.

5.3.2 U.S. Navy

The Navy specifies its flammability requirements for ship interiors in MIL-STD-1623B, "Fire Performance Requirements and Approved Specifications for Interior Finish Materials and Furnishings (Naval Shipboard Use)." (U.S. Navy Sea Systems Command, Washington, D.C. Nov. 14, 1975). The reader is advised to obtain the latest version since this standard is updated on a continuing basis as new data are developed. Also important is the Navy Habitability List (U.S. Navy Habitability Guidance List of Acceptable Materials (Revision D: 1977) available from U.S. Navy, Code 6101, Washington, D.C.).

5.3.2.1 Ships

5.3.2.1.1. Hull Materials

Standard state-of-the-art hulls are of steel construction in accordance with design load criteria identified in the detailed shipbuilding specification. Special lightweight hulls for high performance ships (SES, PHM) may require experimental materials such as new aluminum alloys to achieve structural integrity. In both cases no flammability requirements for hull materials have been applied to date.

5.3.2.1.2 Bulkhead Sheathing

Bulkhead sheathing, when not made of metal may be a high-pressure laminate (for a vertical surface), a fabric-backed vinyl laminate, or a vinyl film-aluminum laminate. In any case, the test procedure required is the ASTM E-84 test, and the flame spread limit allowed is 25. The allowable smoke developed limit is 15 for the first two materials and 75 for the third type. The maximum test limits are based on material bonded to a noncombustible substrate.

5.3.2.1.3 Overhead Sheathing

The overhead sheathing, when not made of metal, is required to be tested by ASTM E-84, with the following maximum limits:

	Flame Spread	Smoke
Fibrous glass opaque suspended ceiling panel	25	35
Acrylic light-duffusing panels windows (lighting fixture only)	250	450
Vinyl film-aluminum laminate, perforated	25	15

The maximum test limits are based on testing materials attached to, or supported by, a noncombustible substrate.

It is noted in this specification that in cases where a flame spread greater than 25 is allowed, it is because no other acceptable material is available. However, polycarbonate light diffuser panels are widely used in the transportation industry; their flame spread rating is significantly less than that of acrylic. Using the ASTM E-162 test method, the flame spread index for polycarbonate is under 100, whereas that for acrylic is approximately 300. It is suggested that this requirement of MIL-STD-1623B be reevaluated.

5.3.2.1.4 Deck Coverings

A variety of polymeric deck covering materials may be specified, including vinyl-asbestos and vinyl tile and sheet, rubber roll, and conductive linoleum. The required test method is Federal Standard 501, Method 6411. This method uses a 30-inch horizontal flue attached to an 18-inch vertical flue. The 31 ¼ inch × 7 inch specimen is mounted horizontally at the base of the horizontal flue. Four burners apply flame to one end of the specimen for 4 minutes. The time from initial application of the flame until flaming ceases (combustion time), is measured, as is the char length.

Depending on the material, the maximum allowable char length is 3 to 10 inches, and for all materials, the combustion time limit is 4 minutes. Further, materials may exhibit ignition time of no less than 30 seconds. Smoke is noted as light, medium, or heavy.

The flammability test appears to be adequate for this category of material. However, the new Flooring Radiant Panel Test, NFPA Standard No. 253-78, should be considered as an alternative. Because of the wide use of these materials, a smoke limit should be applied, as tested by a procedure such as the NFPA No. 258.

5.3.2.1.5 Vessel Furnishing and Decorative Elements

a. Upholstery

All upholstery is tested to comply with Federal Standard 191, Method 5903. This is a vertical flammability test in which a flame is applied to the base of the material for 12 seconds. The average burn length may not exceed 8 inches, and the average duration of flaming after removal of the flame source may not exceed 15 seconds. Drippings may not continue to flame for more than 5 seconds after falling. The Navy specification reduces the limits of char length to 3 inches for vinyl and 5 inches for aromatic polyamide and treated cotton ticking. The afterflame is limited to 2 seconds for all but the aromatic polyamide upholstery, which is 1 second.

The Navy specification has no requirement for smoke emission. This is an important omission. Furthermore, flaming drippings should not be allowed. Smoke emission should be restricted to D of 100 in not less than 4 minutes by the NFPA No. 258 test method.

b. Draperies and Curtains

The test method applied to this category is the same as for upholstery (5.3.2.1.5a), and three materials are specified. A smoke emission test using the NBS Smoke Chamber (NFPA No. 258-76) is also required. The maximum test limits are as follows:

Material	Char Length	After Flame	After Glow	D_s (corr.)
Fibrous glass	1.5 in.	1 sec	5 sec	20
Polyaramid	5 in.	1 sec	5 sec	20
Polyaramid/Novoloid	3 in.	1 sec	5 sec	20

c. Cushioning and Mattresses

The ASTM E-162 Radiant Panel Test is required for this category. The maximum permitted Flame Spread Index is 10. This eliminates the use of polyurethane foam and effectively limits the present choice to neoprene foam. No smoke test is required, because early versions of neoprene foam could not meet smoke emission requirements. However, because of the other favorable fire characteristics of neoprene, trade-off study led to acceptance. A new neoprene formulation has exhibited a D of approximately 140 in 4 minutes; this would be much more acceptable.

d. High-Pressure Laminate for Table Tops

The ASTM E-84 is the test procedure required here. The Flame Spread Limit is 50.

5.3.2.1.6 Thermal Insulation

Most of the thermal insulations are inorganic and are required to pass the U.S. Coast Guard Specification 164.009 test for noncombustible materials.

Pipe and block insulation are tested by ASTM E-84, with flame spread and smoke developed limits of zero.

A PVC-nitrile insulation is also subject to ASTM E-84; it is limited to a flame spread of 25. When tested for smoke by the NBS Smoke Chamber, the D_s limit is 250.

Urethane foam may be used only in reefer spaces, but in such use, must be sandwiched between steel plates. No test specifications are available.

5.3.2.1.7 Acoustic Materials

As of November 1975, no flammability specifications had been adopted for acoustic materials, although many of them contain polymeric materials in part. These acoustic materials are often covered by perforated metal sheathing. This could be a serious omission, depending on location and quantity used.

5.3.2.1.8 Electrical Insulation

The Navy Specification for cable uses MIL-STD-C-915 E, particularly Sections 4.8.16 and 4.8.17 of that specification.

5.3.2.2 Small Boats

The specifications for building a 50-foot utility boat (NS 0902-017-8010), require that general purpose resins shall be of the fire retardant non-air-inhibited type, conforming to Class A of MIL-R-21607. This specification refers to Federal Standard 406, Method 2023. Section 5.3.1.1.1 of this chapter discusses this method, and notes that it is expected to be abandoned in favor of a modified ISO (International Organization for Standardization) test now under investigation by the National Bureau of Standards (Section 5.3.1.1.5).

Other small boats are generally built in accordance with the above same specifications. There are no specifications for other combustible materials.

5.3.3 Smoke Emission

The smoke hazard is further addressed in Chapter 6 of this volume, and in Volume 3 of this series.

Tests for smoke emission are included in flammability test standards ASTM E-84 and E-162, among others. The test procedure that has found greatest acceptance is the National Bureau of Standards Smoke Chamber, which has been formalized as the National Fire Protection Association (NFPA) Standard No. 258-76, "Smoke Generated by Solid Materials." In this method, a 3 inch X 3 inch sample is exposed to radiant heat at 2.5 watts/cm (smoldering mode), and to a series of six micro burners (flaming mode). The smoke density is measured vertically by a photo-electric cell. The measured specific density, D_s, is a function of the chamber parameters. At $D_s = 16$, visibility is approximately 84 percent of original transmission. At $D_s = 100$, visibility is 17 percent, and at $D_s = 200$, it is only 3 percent of original transmission. Correlation with results in room tests has been reasonably good.

Specifications in land transportation vehicles and aircraft require that the smoke emission not exceed D of 100 in 90 seconds, and be no greater than 200 at the end of 4 minutes. The specification derives from the time necessary to evacuate a vehicle. Navy and Coast Guard requirements for smoke emission should relate to requirements for maintaining operations and for firefighting.

5.3.4 Toxicity of Combustion Products

Chemical analysis by itself provides only a limited description of the toxicity hazard of the products of combustion. By this method alone, many organic toxicants would escape detection. A definitive test for toxicity must include an animal test that determines incapacitation and death in the animals and its correlation to a similar effect in humans.

Work now being performed at several centers, including the FAA Civil Aeromedical Institute and the University of Utah, shows considerable promise in the pursuit of a standard toxicity test. The agreement among replicate tests of time to incapacitation and time to death on exposure of rats to gaseous products of combustion is excellent.

Toxicity guidelines must be developed before interior materials toxicity requirements can be specified. An extensive discussion is presented in Volume 3 of this report.

5.3.5 Corrosion

There are no current tests for corrosion potential of gaseous products of combustion of materials. (See Section 5.4.8).

5.4 Discussion of Critical Elements in Flammability Testing

5.4.1 Test Geometry

For materials that are used in a variety of different orientations, a simple horizontal test is wholly inadequate. It is well known that the flame propagation rate in a vertically oriented sample may be more than 10 times that of the same material burning in a horizontal position. The ASTM D-1692 test therefore lacks significance. (See discussion in Section 5.3.1.1.4.b).

Test procedures should, as much as possible, recognize the usual orientation of a material in service. For example, it has been common practice to test carpets using the ASTM E-84 Tunnel Test, which places the carpet on the roof of a tunnel while heating from below. The new NFPA Flooring Radiant Panel Test No. 253-78, more properly places the carpet in a horizontal position so that it is heated from above.

5.4.2 Allowable Flaming After Exposure

Flame time after removal of the ignition source should be limited to no more than 10 seconds and less where possible, so as to provide a minimum of opportunity for ignition of adjacent materials. Flaming drippings clearly represent an opportunity for flame spread, and should not be permitted.

5.4.3 Test Conditions

Flame spread rates and ignition times are usually determined on single, homogeneous specimens. Thus, they may not represent the response of a component made up of several materials in a natural fire environment. It is, therefore, wise to test a multi-material component, such as an entire seat, if possible. Unfortunately, there are no standard tests for such components, so the method of ignition must be left up to the experimenter. In addition to tests on individual materials, a testing scenario should be developed that will contain the ingredients of the ignition and fire propagation observed in previous real fires with similar materials.

5.4.4 Limiting Oxygen Index

The limiting oxygen index test is a useful laboratory test to compare materials in a given category. It can be used to screen out unsatisfactory materials, but should not be used to predict material behavior in a real fire situation.

5.4.5 Rate of Heat Release

Rate of heat release may be defined as the heat produced by the combustion of a given weight or volume of material over a given period of time. This characteristic is relevant to fires because a material which burns with relatively little heat per unit time will contribute appreciably less toward propagating a fire than a material which generates large amounts of heat. Over the past several years there has been growing acceptance among those working in the fire field that this is an important criterion by which to evaluate the fire hazard of a particular material. The rate of heat release is distinct from ignitibility and surface flame-spread potential. Total heat release is another but different qualifying parameter.

No ship specifications at the present time provide for any standards or specifications for heat release. Such methods are now being developed at the National Bureau of Standards and elsewhere.

A description of heat release rate calorimeters is presented in Chapter 5, Volume 2 of this series.

5.4.6 Smoke Evolution

Smoke density may be defined as the degree of light or sight obscuration produced by smoke from burning or pyrolyzing material in a given condition of exposure. This characteristic is relevant to fire safety because escape from a hazardous area is enhanced if the occupant can see the exit and is not incapacitated by smoke constituents.

Two measures of smoke density are the degree of light absorption and the specific optical density.

As noted earlier, the NBS Smoke Density Chamber is the best equipment currently available for the measurement of smoke density, and had been adopted by the National Fire Protection Association (NFPA) as a national standard (NFPA No. 258). Current practice shows a tendency for standardization at a single heat flux level, (i.e., 2.5 watt/cm). Several laboratories are now conducting work to determine the effect of higher heat flux levels. As expected, it has been determined that the maximum smoke density (D_s) can be greater and will be achieved earlier when higher heat fluxes are used.

A National Bureau of Standards report (Breden and Meisters, 1976) describes the effect of heating thermoplastic materials in a horizontal position rather than vertically. Dramatic increases in smoke levels were observed. Materials that do not melt and flow are little affected by this change in orientation.

Other test methods such as ASTM E-84 and ASTM E-162 provide means for measuring smoke, but these are being displaced in favor of the NBS Smoke Chamber.

5.4.7 Toxicity

Numerous test methods for toxicity of combustion products have been devised over the past several years. Chemical analysis by means of standard wet chemistry is augmented by the use of the mass spectrograph and gas chromatography. The resolution of toxicants by these methods can often be laborious and inconclusive.

Live animal tests for toxicants are more conclusive. The relationship of rat incapacitation and death to similar effects in humans has been postulated. Many different methods have been devised to conduct these tests, varying in degree of immersion of the animals and in the degree of sophistication in instrumentation to observe the physiological effects.

A standard test for toxicity is needed; it may be close to realization at this time.

5.4.8 Corrosion

Some of the products of combustion, particularly hydrogen chloride, are extremely corrosive to exposed metal surfaces. Even a small fire in a control room, for example, can result in the necessity to replace sensitive exposed metal parts such as electrical connectors. Consideration must be given to the elimination of materials that can produce corrosive gases from areas that contain sensitive parts.

5.4.9 Fire Endurance

Fire endurance may be defined as the resistance offered by a material to the thermal effects of fire. Two measures of fire endurance are penetration time and resistance time. The use of the ASTM E-119 standard test for the determination of the fire endurance is now widely recognized. For example, the Coast Guard uses the test for deck covering, bulkhead panels, and structural insulation.

5.4.10 Combustible Gas Evolution

Combustible gas from burning or pyrolyzing materials may accumulate and produce flashover under certain conditions. This phenomenon has been observed, for example, where polyurethane foam or foamed latex was used. The low temperature of thermal decomposition permits the production of large quantities of combustible gases by radiant heating.

There is no test which completely defines combustible gas evolution.

5.4.11 Ease of Suppression

Ease of suppression may be defined as the relative ability with which the burning material can be extinguished by a particular extinguishing agent. Extinguishment methods formerly used water. Where this would not be practical, or might cause severe damage to electrical and electronic equipment, flooding with nitrogen, carbon dioxide, or organic halogen gases may be used. Of these, Halon 1301 (CF_3 Br) is often used, particularly where humans may be exposed. If the fire is a surface fire, Halon 1301, acting by a chemical free radical mechanism, will extinguish the fire within 1 or 2 seconds. Deep-seated fires will require a "soaking" period with this extinguishant. Flooding a burning area with carbon dioxide or nitrogen is often used where humans are not expected to be present. In contrast to Halon 1301, carbon dioxide and nitrogen act by limiting the oxygen available to support a fire, and will cause asphyxiation in humans.

Currently there are no standard test methods for ease of extinguishment.

5.4.12 Prediction of Actual Fire Behavior

Prediction of actual fire behavior is made difficult by the large number of possible scenarios that can be conceived for the multitude of ship designs and usage. However, based on a knowledge of material behavior in relevant laboratory tests, and the design of the vessels, an experienced investigator can make a likely prediction of the outcome of a fire. The success of the prediction is enhanced by using results of component testing in large-scale compartmentalized tests.

5.4.13 Testing Adequacy for Systems

A recent development in total system evaluation is an attempt (for aircraft by the FAA) to define a total hazard index that would enable formulation of an equation for a system that could combine flammability, smoke, and toxicity hazards. This concept may be difficult to implement, in part because a fully adequate definition of the specific hazards does not exist. The number and scope of the variables in any system presents a formidable challenge. How much weighting should be given to each of the variables? Will the variables and weightings differ substantially from vessel to vessel? These and other questions must be answered before credence can be given to this type of evaluation.

5.4.14 Modeling and Scaling

Techniques for modeling and scaling for fire hazard are not now sufficient to justify reliance on these techniques for ships.

Efforts to perform mathematical modeling of systems have not been successful because of the large number of variables and because the basic predictability of the rates of heat and smoke release and flame spread are affected by many uncontrollable factors.

Scaling also has been found to be largely unproductive, although recent scaling work in which the system is placed under pressure has achieved some success on simple models.

5.4.15 Large-Scale Testing

Large-scale testing, as distinguished from full-scale testing, is performed by segmenting an entire vessel to permit sections to be tested under conditions simulating a real fire. In performing these tests, one can determine the interaction of several different types of materials that are normally positioned adjacent in the structure of a vessel.

It is essential that the real environment affecting the components be duplicated as closely as possible, including air convection and radiation effects. Furthermore, the volume in which the test is conducted should not constrict the fire any more than would be encountered in a full-scale fire situation.

The Coast Guard conducts such large-scale testing at its Fire and Safety Test Facility at Mobile, Ala.

5.4.16 Full-Scale Testing

Fire testing of a total vessel is uneconomical except, perhaps, for small boats. Furthermore, it does not permit repeated testing using different scenarios for ignition, or the use of a variety of different materials or components.

5.5 Programs Needed for Improved Specifications

The Coast Guard and the Navy continually review their specifications. This involves the updating of specifications and the establishment of specifications for applications not previously covered. Both agencies are seeking outside assistance in the revision of the specifications. In May 1977, the Navy sponsored a National Materials Advisory Board symposium and workshop to review standards for flammability and smoke emission of electrical insulation for wire and cable. The Coast Guard has asked the National Bureau of Standards to develop a new standard for flame propagation.

The need for a standard for measuring the toxicity of gaseous thermal decomposition products is recognized as a requirement for characterizing ship materials and for other applications. Using such a standard, appropriate specifications can be established.

The Navy has described the current situation that has increased the fire hazard in its ships, and has documented its requirements. The seriousness of the fire threat has become more significant in recent years. The use of combustible plastics and other organic materials contributes significantly to the fire load. Even materials which carry the label, "self-extinguishing," or "flame retardant" often contribute significant quantities of toxic gases and evolve dense, vision-obscuring smoke when exposed to a fire. A derivative problem is the sensitivity of today's weapons and weapon systems to damage by fire and combustion products. The complexity and sophistication of advanced weapon systems require that all components function in time of stress. A fire that causes abandonment of vital spaces or one that destroys vital electronics components and electrical cables severely compromises combat capability.

In addition to a program for the development of new materials, a major effort is needed to develop test methods to permit a realistic prediction of material behavior in fire situations of the types likely to be encountered on a naval vessel. Also required is a definitive means of evaluating pyrolysis and out-gassing effects of shipboard materials.

5.6 The Adequacy of Materials Testing

Flammability test methods should have some sort of commonality in describing the fire response of potentially combustible materials over a wide range of applications. However, misunderstandings and improper evaluations can occur unless the limitations of the test methods are understood.

To the uninitiated, it would appear that a material which exhibits superior performance in one fire test would exhibit superior performance in any other. Unfortunately, this is not the case. Relative performance can vary considerably, depending upon the test used.

A single fire test cannot predict the behavior of a material under all possible conditions of fire exposure. A combination of fire tests can come closer to such a prediction. Economic factors alone usually prohibit the performance of all fire tests that could be relevant to all possible fire scenarios; therefore, fire tests can and should be classified according to what is more desired to be learned about a given material, component, or system.

Fire tests can be divided into three groups:

1. Laboratory research and development tests designed to generate information on the basic properties of a material, or combination of materials, and the effects of different variables on those properties. Such basic measurements would include ignition temperature and heat of combustion.

2. Test procedures designed to simulate anticipated application conditions and intended to serve as standards on which specifications may be based. Examples are the Federal standards for carpets, mattresses, upholstered furniture, etc., that use such test procedures.

3. Large-scale or full-scale tests designed to reproduce actual fire scenarios under controlled and measured conditions. Such tests can provide a realistic basis for judging the validity of the pragmatic tests used as standards for specifications.

Because of economic considerations, relatively few full-scale tests have been performed to validate the standard simulation tests. Regulatory agencies have, to a great extent, relied on previous fire experience to select the tests needed for their particular situations. Although the test methods used by the Coast Guard and Navy are generally the best that state-of-the-art testing provides, they are inadequate to provide complete guidance for the selection of polymeric materials to be used in systems which may face a wide spectrum of fire scenarios. Many of the test methods are, at best, adequate only for eliminating individual materials whose fire characteristics are blatantly unsatisfactory.

5.7 Conclusions and Recommendations

Conclusion: The use of ASTM D-1692 test for flammability of furniture is not recognized as an acceptable test by the fire community. *Recommendation:* Consider the modified ISO test method now under development to replace several of the Coast Guard flammability tests and as an eventual replacement for ASTM D-1692. In the interim, test fabrics by Federal Standard 191, Method 5903, and other polymeric materials by ASTM E-162, with appropriate limits of flame spread.

Conclusion: The testing of carpets by the ASTM E-84 test method is no longer considered a valid procedure for carpets. *Recommendation:* Use the new NBS Flooring Radiant Panel test, NFPA Standard No. 253-78, in place of E-84 as a flammability specification for carpeting and floor coverings.

Conclusion: The Coast Guard has no flammability specification for polymeric materials used in small boats despite the particular susceptibility of pleasure craft to fires because of inexperienced crews and poorly safeguarded fuel sytems. *Recommendation:* Consider, when development is completed, the modified ISO test methods, with appropriate limits, for application to polymeric materials used in small boat construction. In the interim, preparation should be made to monitor the development of the ISO method or other acceptable test methods to achieve a series of specifications for the various categories of materials.

Conclusion: Fuel systems are not a part of the Committee's charge, the design of fuel systems in small boats clearly needs additional consideration to improve safety. Although the Coast Guard and Navy use tests that are presumably adequate, many of the test methods available to them are inadequate to provide complete guidance for the selection of polymeric materials to be used in systems which might face a wide spectrum of fire scenarios. *Recommendation:* Actively pursue current efforts by the Navy and the Coast Guard to monitor test method development, to adopt test methods in areas where none are currently specified, and to develop new test procedures.

Conclusion: The Navy specifications have no requirement for smoke emission from upholstery materials which are frequently heavy contributors of smoke. In addition, the specification provides that flaming drippings may continue to burn for 5 seconds. *Recommendation:* Establish a Navy smoke emission standard for upholstery, in which the specific density (D_s) should not exceed 100 in less than 4 minutes by the NFPA No. 258 test method. Change the specification to forbid flaming drippings.

REFERENCE

Breden, L. and Meisters, M., NBSIR 76-1030, National Bureau of Standards, Washington, D.C., February 1976.

CHAPTER 6

SMOKE AND TOXICITY

6.1 Introduction

In general, the polymeric materials and the environment found on board naval, commercial, and recreational vessels are similar. It follows that the smoke and toxicity problems arising as a result of the combustion or pyrolysis of these polymeric materials will be similar, except under military combat conditions.

In Volume 3 of this series, the committee discussed the nature of the products of combustion or pyrolysis of polymers and the types of toxicity that can result when humans are exposed to those products.

Ships and boats are confined spaces in which fire represents a serious hazard. Usually they operate as self-contained vessels, from which escape may not be feasible, and are remote from outside firefighting assistance. Although ships are largely constructed of metal, significant amounts of polymers are used in interior finish materials, furnishings, communications, and control systems. In contrast, polymers are widely used as structural materials in recreational boats. The materials on ships and boats that are most likely to undergo combustion or pyrolysis (excluding, for the purpose of this report, engine fuel, lubricants, and hydraulic fluids) are natural and synthetic polymers. With the exception of wood, cotton, paper, and other cellulosic materials, the large majority of such flammable materials on ships are synthetic polymeric materials.

Polymeric materials in use aboard ships and boats cover a broad spectrum (see Chapter 4). Although there are prescribed tests for other flammability characteristics of polymers used aboard ships and boats, and limited specifications relating to smoke emission for commercial and naval vessels, there are no specifications to be met for toxic gas generation under fire conditions. Survival from fire in confined spaces is dependent upon thermal energy flux, oxygen concentration, toxic gases, smoke, fear or panic, and concurrent psychophysiological disease or impairment, each of which is discussed in more detail below.

6.1.1 Thermal Energy

The human hazards are either local thermal damage, usually of the skin or respiratory airway, or generalized thermal shock. Naval ships are equipped with sophisticated firefighting equipment. In a fire, crew members normally put on oxygen masks and fight the fire in their assigned areas. In essence, they become firefighters. As with firefighters, thermal injury is an occupational hazard and a significant cause of death among naval crew members in shipboard fires. The same statement is not necessarily true for crew members or passengers aboard commercial vessels who are not required to fight fires.

6.1.2 Decreased Atmospheric Oxygen Concentration

Decreased oxygen is associated with the rapid material oxidation that occurs in fires. The health hazard is deprivation of an essential amount of respirable oxygen.

6.1.3 Carbon Monoxide and Other Toxic Gases and Aerosols

Incomplete combustion of polymers in the presence of oxygen yields carbon monoxide. The consequences of exposure to carbon monoxide have been extensively reviewed (National Advisory Center on Toxicology, 1973; Stewart, 1976). Carbon monoxide inhalation is the major cause of death in fires. In addition to carbon monoxide, pyrolysis or combustion of polymers may produce combinations of toxicants including hydrogen cyanide, sulfur compounds, nitrogen oxides, halogen acid gases, organic halide gases (e.g., ethyl bromide), and aldehydes (e.g., acrolein). Fire suppressant additives in the polymers may produce additional toxic products when the polymer undergoes combustion.

6.1.4 Smoke

Massive exposure to smoke may result in direct mechanical plugging of airways in the lungs and pharynx. Smoke particles may carry adsorbed gases, liquids, or residual heat to various parts of the respiratory tract. Smoke also obscures vision, impedes escape or rescue, and hinders firefighting. Irritants in smoke may cause temporary blindness, reflex coughing, and impede escape, as well as hinder firefighting.

6.1.5 Fear, Panic, or Incidental Trauma

Fear, panic, or incidental trauma are recognized causes of incapacitation in fires. Fear or panic can cause inappropriate reaction and result in trauma, for example by trampling, as hysterical people rush for an exit in an attempt to escape from a smokefilled room.

6.1.6 Pre-existing or Concurrent Psychophysiological Disease or Impairment

Pre-existing heart disease is a significant factor in fire deaths. Alcohol intoxication may be a cause of delayed fire detection, or inability to escape. Escape and survival are especially difficult for physically handicapped persons. Other diseases and impairments may also limit an individual's ability to escape and survive.

6.2 Perspective on Toxicity and Smoke

In the past decade, there have been an increasing number of studies of toxic gases and smoke resulting from pyrolysis of polymeric materials.

Volatile pyrolysis or combustion products of polymers used in ships or boats include carbon monoxide (CO), carbon dioxide (CO_2), hydrogen cyanide (HCN), oxides of nitrogen (NO_x), ammonia (NH_3), hydrogen sulfide (H_2S), phosgene ($COCl_2$), aldehydes, and many other compounds. Each of these gases adversely affects one or more parts of the respiratory function of the body. CO interferes with the oxygen transport in the blood; high concentrations of CO_2 depress respiration; HCN inactivates the respiratory enzymes in the tissue; H_2S produces respiratory paralysis; and ammonia, phosgene, aldehydes, and the oxides of nitrogen produce irritation in the respiratory tract. Additional details regarding the toxicologic effects of these gases are described by Wagner (1972) and Yuill (1974). Kimmerle (1974) gives a detailed compilation of effects at various atmospheric concentrations.

The predominant toxic polymeric combustion product is CO. Incapacitating or lethal amounts of CO can develop within minutes. A brief review of CO toxicity was presented by Montgomery, Reinhardt, and Terrill (1974). The net physiological response to CO and other thermal degradation gases is far from clear, although awareness of the problem and the difficulties of properly assessing it is increasing. The toxicology of fires can be extremely complex. If a lethal CO atmosphere is not reached, other lethal or disabling factors may still be present. For example, in the series of experiments reported by Cornish and Abar (1969), pulmonary injury from HCl developed in the absence of lethal effects from CO.

An important effect noted by Effenberger (1972) is that under certain conditions burning polystyrene does not cause rats to develop significant amounts of carboxyhemoglobin, but rather, the resulting styrene monomer apparently immobilized the rats. The pyrolysis of polystyrene is merely one example of a specific product causing incapacitation. Other combustion products may cause varying degrees of incapacitation, resulting in serious injury or death. If such interpretations may be extended to humans exposed to fires, death could result because incapacitation reduced one's ability to escape from the fire.

The composition of smoke varies widely, and smoke holds many hazards. Smoke is basically a mixture of carbon particles and other solids resulting from combustion. It may contain irritants adsorbed on the particles and be mixed with thermal decomposition gases. The hazards of smoke may be physical (blocking vision or airways), physiological (local or systemic chemical irritation), toxicological, thermal (heat injury), or psychological (fear and panic). Gaskill (1973) surveyed the hazards from smoke and described contemporary measurement techniques.

Because the factors affecting toxicity are so complex and the data often unavailable or conflicting, the reader is urged to review the references in this chapter, and Volume 3 (Smoke and Toxicity) of this series to obtain a more complete picture of the current understanding of toxicity of combustion products of polymers.

One of the basic problems relating to toxicity is the inadequacy of the data base. Although there have been many studies and much data gathered, there has not been adequate codification and evaluation. In Volume 6 of this series, the committee recommends that a central agency be designated to collect and analyze data relating to the fire safety aspects of polymeric materials, including toxicity and smoke. This recommendation is reiterated and strongly endorsed.

6.2.1 Research and Development

The fire threat aboard naval ships and craft has become more serious in recent years. The use of combustible plastics and other organic materials in the forms of electrical cable insulation, more modern furnishings and thermal hull insulation contributes significantly to the fuel load and, thus, to the hazard. The need exists to provide materials which exhibit improved fire performance characteristics, including reduced flame spread, less vision-obscuring smoke, and production of less toxic substances.

Present-day efforts to minimize the shipboard fire has largely been an outgrowth of the Habitability Improvement Program. Prior to 1965, ships in service were constructed to habitability standards of pre-World War II vintage. Items such as plush carpets, overstuffed furniture, and a variety of wooden furnishings were installed by chance rather than by design. Introduction of the "self-help" concept created a surge in upgrading shipboard habitability. This bootstrap process proceeded without giving much consideration to the flammability aspects of materials. Unfortunately, the installation of combustibles with no concurrent improvement in fire protection systems resulted in degradation of mission capability and ship safety.

The multi-million-dollar fire aboard the USS Forrestal in 1972 (primarily involving so-called "habitability materials") identified the need to reestablish control of shipboard combustible materials and eliminate unnecessary furnishings. Orders to rip-out were received with reluctance, especially when replacement materials were not readily available. Several questions were immediately apparent:

1. How does one determine the basis for acceptance or rejection of materials?
2. What is considered unnecessary?
3. How will the policy be implemented and enforced?

The approach to resolving these questions was implemented by the Chief of Naval Operations through the Chief of Naval Material. The plan included a comprehensive test and evaluation program for assessing the fire performance characteristics of materials, and ultimately led to the development and promulgation of MIL-STD-1623. This document established fire performance requirements and identified nearly 50 approved specifications for the following seven categories of interior finish materials and furnishings used on new and existing surface ships and submarinees:

1. Bulkhead sheathing
2. Overhead sheathing
3. Deck coverings
4. Furniture
5. Draperies/curtains
6. Thermal insulation
7. Acoustic materials

Major items considered in determining the fire safety aspects of materials contained in MIL-STD-1623 include surface flammability, vertical flame resistance, smoke generation, and tests for incombustibility. It is recognized that the development of limits for toxic products of combustion would provide a further benefit in assessing the fire hazard of materials in question. The information generally available, however, is not refined to the degree that will now allow inclusion of finite limits in MIL-STD-1623. This document is implemented by the applicable material specifications and is enforced by individual ship building overhaul contracts. It is planned to incorporate additional fire criteria in the standard as the required data becomes available.

The National Bureau of Standards (NBS) has been involved in flammability research into materials used in crew compartments. This effort has resulted in the development of novel small-scale fire tests and associated fire criteria. Recommendations are being reviewed for possible inclusion in MID-STD-1623.

The Naval Sea Systems Command (NAVSEA) has sponsored several other research and development programs intended to reduce the shipboard smoke and toxicity fire hazard associated with organic materials. Polyphosphazene (PN) polymers are being investigated for a variety of applications, including electrical cable insulation, hull insulation, and coatings. These nonhalogenate polymers are more difficult to ignite, release smaller quantities of toxic products, and tend to evolve less-vision-obscuring smoke under fire conditions than other materials used in similar applications.

The application of fire barriers to existing material installations have resulted in significant reductions in fire risk. NAVSEA has asked NBS to evaluate intumescent coatings as a means of providing a thermal barrier for use on hull insulation. Full-scale compartment fire tests have demonstrated the feasibility of protecting combustible substances by intumescent coatings; evaluation on board selected ships is now underway.

Modern electrical cables employ organic polymers as insulation and jacketing materials; these polymers present a major problem during a fire. While the cables are generally not the source of fire ignition, they are highly vulnerable and once ignited will burn and rapidly spread fire. In addition, these polymeric materials generate dense smoke and toxic/corrosive products of combustion which severely limit firefighting capability, threaten the life and health of those exposed, and significantly increase fire damage costs and repair time. Efforts are underway to investigate material and application techniques for new and existing ships to reduce the involvement of electrical cables. Some progress has been made by use of a sprayable inorganic fiber/binder system which has shown adaptability to the variety of cable configurations existing aboard ship.

With regard to smoke, the Navy is exploring a polymer additive approach to minimize vision-obscuring particulate matter generated by burning electrical cables and cushioning. Initial efforts have produced significant smoke reduction in polyvinyl chloride cable jacketing material with no apparent degradation of electrical properties.

Making a product "fire safe" is a complex problem with many interactive variables. The current "design to cost" concept requires a look toward more cost-effective means of addressing and reducing shipboard fire risk through more intensive materials research and development efforts.

6.2.2 Clinical Data

Quantitative toxicity data derived from naval, commercial, and recreational vessel fire victims are non-existent; however, limited data are available from aircraft fires and are pertinent, since the polymeric materials involved and their system applications are similar.

Smith and his associates (1976) have described the results of their forensic investigations of several aircraft accidents (see National Research Council, 1979). Two commercial aircraft accidents in the United States (Denver, Colo., 1961; Salt Lake City, Utah, 1965) contributed greatly to the present concern about the toxic hazard of gases generated in aircraft fires. Analysis of these accidents showed that most of the occupants probably suffered relatively mild impact trauma; yet, 60 persons died from thermal and chemical injuries in the ensuing fires. Carboxyhemoglobin measurements on 16 victims of the Denver crash revealed CO saturations ranging from 30 to 85 percent (the mean was 63.3). Similar analysis on 36 victims of the Salt Lake City accident indicated CO saturations ranging from 13 to 82 percent (the mean was 36.9). The lower carboxyhemoglobin values found in the second accident have been attributed to the presence of fire within the aircraft before evacuation could be attempted and that many victims must have died quickly of direct thermal effects. Because of the chemical nature of the polymeric materials involved, it is assumed that gases other than carbon monoxide contributed to the toxicity of the cabin environment, but there is no supporting evidence for this assumption. In 1970, blood samples from victims of an aircraft crash followed by fire (Anchorage, Alaska, November 1970) were analyzed for the presence of cyanide. This was the first time, to the best of our knowledge, that such analyses had been made on victims of an aircraft fire. Measurable amounts of cyanide were found in 18 of 19 specimens; the accompanying carbon monoxide saturations ranged from 17 to 70 percent. In the one sample in which cyanide could not be detected, the carboxyhemoglobin concentration, 4.9 percent, did not exceed that which could result from smoking, indicating the probability of death on impact. Blood cyanide levels in these victims corresponded closely with those reported in the literature for victims of structural and vehicular fires, ranging from the lower detection limit (ca. 0.01 microgram/ml) to 2.26 microgram/ml. The relationship between cyanide levels and carboxyhemoglobin content varied in random fashion, perhaps representing relative proximity of the victims to cyanide-producing materials. Alternatively, the varying cyanide levels reported may be due to uncontrolled auto-production of cyanide in and from the tissues. Nothing in these findings permitted speculation concerning the relative contribution of the two gases to the deaths. In addition, there was no way of assessing the possible contribution of other gases that must have been present in the pyrolysis products to which these victims were exposed.

6.3 Evaluating the Hazards of Toxic Fumes and Smoke

6.3.1 Comparison of Materials

Valid comparisons of different materials must be based on similar or reasonably standard conditions. The advantages of a standardized toxicity test procedure pertinent to proposed use is obvious. However, the chemical properties and physical state of polymers cannot be used to predict the concentration of toxic products or smoke in real fire situations. Critical properties often are not known or may vary due to production techniques; furthermore, standard fire condition tests cannot be applied for predictive purposes. Real fires possess two essentially uncontrollable variables — oxygen supply and temperature. These make selection of test procedures inherently difficult. Generally, laboratory pyrolysis tests are made in an oxygen-lean atmosphere that will not support combustion, while flame tests are made in an oxygen-rich atmosphere. Since either pyrolysis or combustion can be more hazardous, depending on the nature of the material being consumed, a standard procedure must, of necessity, take both processes into account separately and together. (See Volume 3 of this report for a discussion of methodology). In the case of fires aboard naval vessels, it should be noted that accepted firefighting procedures involve isolation and confinement of fires by closing off compartments with airtight doors. In such cases the fires occur in oxygen-lean atmospheres and high carbon monoxide levels would be expected.

6.3.2 Thermal Decomposition Temperatures

The nature of combustion products can vary significantly, depending upon fire temperature. This, in turn, is dependent upon the pyrolysis and/or combustion process, oxygen concentration, and caloric value of the products consumed by the fire. Although there have been attempts to classify the toxic and gas products from wood by using four distinct temperature zones (Beall and Eickner, 1970), the committee is unaware of any attempt at such a classification of the multitude of synthetic polymers that exist today.

6.3.3 Method of Study

There are four recognized methods for studying fires. They are:

- chemical analysis
- biological testing
- predictive testing
- epidemiological studies

Recent literature describes the four types of methods. (See Einhorn et. al., Reinke and Reinhardt 1973; Kimmerle 1974).

6.3.3.1 Chemical Analysis

Identification of the chemical components of the combustion products of polymers can help in understanding the effect of altering variables such as burning temperature and oxygen supply. The relative hazard or lethality of the product can be estimated with reasonable confidence if a single component, such as CO or HCl, is clearly predominant and no other significant source of physiological stress is present. On the other hand, if thermal degradation generates a significant quantity of diverse gases accompanied by heat and/or smoke, although the overall physiological response may be evident, the contribution of the individual agents is not identifiable. The problem is complicated in that the polymeric product may contain a variety of additives, including anti-oxidants, fillers, stabilizers, and modifiers. In some cases, even the major polymer components are not disclosed to the user because the supplier considers the information to be proprietary.

6.3.3.2 Biological Testing

Data obtained from tests on laboratory animals can be expressed as the LC_{50} (i.e., the concentration expected to produce death in 50 percent of the exposed animals), or it may be expressed as the concentration expected to produce a specified effect such as incapacitation. To this latter area special attention should be directed. Significant incapacitation may occur at much lesser dosages or shorter times of exposure than those that cause death. Yet the incapacitation may be of such a degree that the ability to escape is impaired and death may occur due to thermal injury or incidental trauma.

A safer product from the standpoint of a flammability test does not always result in a safer product in that:

1. Polymer structure modification or additives may indeed result in retarded combustion of the polymer but it may be accompanied by increased toxicity of decomposition gases and/or dense smoke from the smoldering of the polymer when it is exposed to heat from the combustion of other fuels.
2. Many common fire retardants contain halogens, such as chlorine and bromine. These can, under appropriate circumstances, produce thermal decomposition products such as hydrogen chloride, phosgene, and hydrogen bromide.
3. Nitrogen, either in the polymer molecule, or in additives, may yield hydrogen cyanide or nitrogen oxides or both as a result of thermal reactions.
4. Polyurethane polymers based on propoxylated trimethylopropane polyols and fire-retarded with phosphorus-containing retardants may yield highly toxic bicyclic phosphorus esters when thermally degraded (Petajan et. al., 1975).

Because of the foregoing problems, as well as those discussed in Sections 6.3.1 and 6.3.2, it is essential that a test method be used that embodies all of the factors discussed. At present the only method which accomplishes this is one that utilizes laboratory animals. Therefore, it is recommended that the use of tests employing laboratory animals be expanded.

6.3.3.3 Predictive Testing

Predictive testing includes combined testing, "room" or large-scale fire tests, computer model testing, etc. None of these have been determined to predict accurately the expected toxic gas and smoke from burning polymers (see Chapter 5). Nevertheless, small-scale tests under controlled and well-specified conditions can produce comparative information about polymeric materials.

6.3.3.4 Epidemiological Studies

Critical epidemiological analyses have been applied to fire toxicity only recently and have generally been limited in scope and usefulness. For marine transportation, there is a need for carefully analyzed reports from naval, commercial, and recreational vessel fires that provide comprehensive casualty data, including quantitative toxicological and pathological evaluation of victims. At present, such data are too sparse to be useful. There is need for such data as well as need to establish guidelines for their collection.

6.4 Special Toxicity Considerations for Ships

Products or formulations require more stringent evaluation when intended for marine vessel usage than for many other applications. The unavoidable presence of large quantities of highly flammable fuel, the long and devious escape routes from compartments below decks, and the unavailability of nearby havens of safety expose ship or boat occupants to a greater potential fire and toxicity hazard than exists in more ordinary situations. If new products or formulations are to be used in marine vessels in appreciable quantity, the potential toxicity of their combustion products must be evaluated experimentally and the information reviewed in the context of available epidemiological data.

6.5 Conclusions and Recommendations

Conclusion: Thre is a lack of information on the special problems of combustion and pyrolysis, smoke, and the composition of products from the combustion and/or pyrolysis of polymeric materials aboard ships. The performance of a polymeric material under laboratory conditions may differ markedly from that which occurs when it, individually or in combination with other materials, is involved in a real shipboard fire. *Recommendation:* Develop a research program which will enable the acquisition of information relating to smoke and toxicity problems that are specific to shipboard use. Every effort should be made to take advantage of actual shipboard fires for the purpose of acquiring information to enable the improvement of specifications of polymers and enhancing medical care of victims of shipboard fires.

REFERENCES

Beall, F. C. and Eickner, H. W., "Thermal Degradation of Wood Components," U.S. Forest Products Laboratory Report 130, Forest Service, U.S Department of Agriculture (Madison, Wis., May 1970).

Cornish, H. H., and Abar, E. L., Toxicity of Pyrolysis Products of Vinyl Plastics, Archives of Environmental Health 19:1–15, 1969.

Effenberger, E., *Toxische Wirkungen der Verbrennunqsprodukte von Kunstoffen, Stadtehygiene* 12; 275–80, 1972.

Einhorn, I., et al., "Physiological and Toxicological Aspects of Smoke Produced during the Combustion of Polymeric Materials," Proceedings of the NSF/RANN Conference on Fire Research (Washington, D.C.: UTEC-MSE 74-083, FRC/UU-29, June 21, 1974). p. 199.

Reinke, R. E. and Reinhardt, C. F., "Fires, Toxicity, and Plastics," Modern Plastics *50,* pp. 94–95, 97–98 (1973).

Kimmerle, G., *Aspects and Methodology for the Evaluation of Toxicological Parameters During Fire Exposure,* Journal of Fire and Flammability/Combustion Toxicology 1:4–51, 1974.

Montgomery, R. R., Reinhardt, C. F., and Terrill, J. B., "Comments on Fire Toxicity," paper presented at the Polymer Conference Series (Flammability of Materials Program), University of Utah, Salt Lake City, Utah, July 11, 1974.

Gaskill, J. R., "Smoke Hazards and Their Measurement; A Researcher's Viewpoint," J. of Fire and Flammability, *4* pp. 279–298 (1973).

National Research Council, Report of Committee on Fire Toxicology, Washington, D.C. (1976).

National Advisory Center on Toxicology. National Academy of Sciences-National Research Council, Committee on Toxicology. Guides for Short-term Exposures of the Public to Air Pollutants. VI. Guide for Carbon Monoxide. Washington, D.C., March 1973.

National Research Council, "Fire Toxicology: Methods for Evaluation of Toxicity of Pyrolysis and Combustion Products-Report No. 2, "Report of Committee on Fire Toxicology, Washington, D.C., 1977.

Petajan, J. H. and Voorhees, K. J., et al., "Extreme Toxicity from Combustion Products of a Fire-Retarded Polyurethane Foam," *Science, 187* pp. 742–44, (1975).

Smith, P. W., Crane, C. R., Sanders, D., Abbott, J., and Endecott, B., "Material Toxicology Evaluation by Direct Animal Exposure." Presented at the International Symposium on Toxicity and Physiology of Combustion Products, University of Utah, Salt Lake City, Utah, March 22–26, 1976.

Stewart, R. D., "The Effect of Carbon Monoxide on Humans," *Journal of Occupational Medicine, 18:* 304–309, 1976.

CHAPTER 7

SURFACE SHIPS: COMMERCIAL AND MILITARY

7.1 Introduction

For the purposes of this discussion, the term "surface ships" is used to refer to floating vessels of all varieties, both foreign and domestic. The two broad classes of surface vessels are military and commercial. The former includes all vessels whose main functions are combat or combat support (both Naval and Coast Guard) while commercial vessels are those designed primarily for domestic and foreign trade of all varieties, including the transport of passengers or freight and the provision of services. In the latter category are oceanographic vessels, used in the collection of scientific data, and mobile offshore drilling units. Such high-performance vessels as hydrofoils and hovercraft are also divided into military or commercial types, depending upon their function. Although these latter vessels differ considerably in design from conventional floating vessels and are constructed of lightweight materials, their safety requirements have been developed using the same principles as found to apply to displacement type hulls. The fire safety problems of the high performance ships are sufficiently special that they have been placed in a separate chapter.

The appendices referred to in this chapter have been placed at the end of this chapter for convenience.

7.1.1 Commercial Vessels

The control of polymers used in the construction or fitting of commercial vessels, as well as the development and enforcement of safety and construction standards, is the primary responsibility of the U.S. Coast Guard. Using safety concepts defined by the Safety of Life at Sea (SOLAS) International Conventions as guidelines, the Coast Guard prepares regulations for the proper shipboard use of polymers. A discussion reviewing the Safety of Life at Sea Conventions dealing with international acceptance of noncombustible materials used on board surface vessels is included in Appendix A.

The Coast Guard is also charged with the responsibility of assuring safety of life and property on U.S. flag vessels at sea. The goal of the Coast Guard merchant vessel fire protection regulatory program is the "Development of fire protection standards adequate to minimize the incidence and consequence of Shipboard fires consistent with economic consideration." This goal is the direct result of a legislative mandate which delegated the responsibility for the development of fire protection and other safety requirements related to the construction and operation of merchant vessels to the Coast Guard. The construction of commercial vessels is in accordance with Coast Guard approved plans and specifications which comply with the regulations specified in various subchapters of the Code of Federal Regulations, Title 46 — Shipping.

The rules issued by the American Bureau of Shipping (ABS) and designated

"Rules for Building and Clearing Steel Vessels" will, in general, be accepted as standard in reviewing the plans and specifications for the construction, alteration, or repair of vessels. Additional requirements for these plans are contained in the following Title 46 subchapters: D-Tank Vessels (46 CFR Parts 30–40), H-Passenger Vessels (46 CFR Parts 70–89), I-Cargo and Miscellaneous Vessels (46 CFR Parts 90–109), T-Small Passenger Vessels Under 100 GT (46 CFR Parts 175–187), and U-Oceanographic Vessels (46 CFR Parts 188–199). Each of the subchapters also refers to specific requirements in Subchapters F-Marine Engineering (46 CFR Parts 50–64), J-Electrical Engineering (46 CFR Parts 110–113), and Q-Specifications (46 CFR Parts 160–164). These standards are applicable to typical vessels. For a vessel of unusual form, arrangemnent, or construction, at least the same degree of safety must be maintained as is established in these standards. When plans approved by ABS are received, they are, in general, accepted as satisfactory, except insofar as the law or regulations contain requirements which are not covered by the rules of the ABS.

In conjunction with its ships safety responsibility, the U.S. Coast Guard is the technical representative of the United States to a specialized agency of the United Nations, the Intergovernmental Maritime Consultative Organization (IMCO). IMCO is the depository for the 1960 International Convention for Safety of Life at Sea; to which the United States is signatory. A revised 1974 Convention was recently submitted to the U.S. Senate for its advice and consent. This treaty will result in a major upgrading of the fire safety requirements for merchant vessels.

The Maritime Administration becomes involved in the development of detailed specifications for commercial vessels in much the same manner as the Federal Housing Authority requires certain minimum standards before approving a home mortgage. The Maritime Administration is the federal mortgage insurer for vessels. The minimum design standards relating to polymer use, utilized by both the U.S. Coast Guard and the Maritime Administration are discussed in detail in subsequent sections.

While there are certain materials that have been used consistently in maritime applications, the influx of manmade polymeric materials as replacements for natural polymers and metals in land-based industrial applications and structures has been somewhat paralleled in the marine industry.

The fire safety of polymeric materials must be examined in view of the under-lying marine fire protection philosophy which evolved as a result of catastrophes like the SS Morro Castle disaster in the 1930s. Implementation of this philosophy results in the following guidelines, as given in Chapter II-2 of SOLAS 74.

1. Division of a ship into main vertical zones by thermal and structural boundaries.
2. Separation of the accommodation spaces from the remainder of the ship by thermal and structural boundaries.
3. Restricted use of combustible materials.

4. Detection of any fire in the zone of origin.

5. Containment and extinction of any fire in the space of origin.

6. Protection of means of escape and access for firefighting.

7. Ready availability of fire extinguishing appliances.

8. Minimization of possibility of ignition of inflammable cargo vapor.

These philosophical guidelines have resulted in regulations which restrict the marine use of plastics from a fire safety standpoint.

7.1.1.1 Automatic Detection

When ships were not too large, visual detection of fires by crewmen or passengers was reasonably effective, but with larger passenger ships, some automatic detectors are necessary. It is important that the characteristics of detection devices be determined before they are fitted to the vessel.

7.1.1.2 Extinguishers

In most ships, the engineering spaces have automatic or remotely operated extinguishment devices. In the passenger compartments, extinguishment generally is manual. The firefighting agent may be water, CO_2, Halon, or other compounds. The burning characteristics of the polymers involved and the size of the fire will largely determine the effectiveness of the firefighting effort. Fires in the passenger compartments of most ships, heavily loaded with polyurethane material (in a variety of applications) as well as polymeric panels, furnishings, etc., may be extremely difficult to extinguish unless the fire is discovered early (i.e., in 1 to 5 minutes) (see 3.3.1.7 and 3.4.1.7).

Flashover (discussed in more detail in Chapter 3, Section 3.3.1.8 and 3.4.1.8) is a critical turning point in shipboard fires, since it is probable that its occurrence signals complete destruction of the compartment, possible loss of life or severe injury to passengers, and an increased likelihood of the fire spreading to adjoining areas. Ship configurations, amount and availability of polymer fuel, and ship operating conditions (good ventilation, etc.) bear importantly on flashover. This is especially so where there is low probability of adequate nearby fire suppression (i.e., where firefighting manpower and response time may be less than adequate). Therefore, ship design must be modified by use of suitable extinguisher systems to protect the passengers and reduce unnecessary losses. (See Sections 3.3.1.8 and 3.4.1.8).

Interiors of ships contain many polymers, all of which contribute to the fire load. (See Section 7.4.1.2 and Table 7.4-5 for materials used). None are used as extensively, or contribute so heavily to fire susceptibility, as urethane foam which is in general use in tanks, reefer spaces, pipe covering, etc. This polymer should be replaced or protected by a noncombustible covering because it is easy to ignite, has a high heat release rate, and contributes large quantities of heavy smoke and toxic gases.

Under normal conditions, passenger egress from staterooms to large open areas is reasonably good for the physically able. In a rapidly developing, smoky fire, egress through doors is more difficult; it is particularly so for the aged and handicapped. Ship design can provide good escape routes (when all regulatory requirements are considered). Improvement in passenger safety should take the form of increasing the time available for egress by reducing the probability of ignition and slowing the spread of fire.

7.1.2 Naval Vessels

The proliferation of polymer usage aboard naval vessels has been slow because of the complex nature of ships, the variety of their missions, and the many materials which have failed to meet existing flammability requirements. Despite this, polymers have found relatively wide use in the habitable areas of many naval ships because of their economic and aesthetic advantages. Polymeric materials used in such areas are addressed in the "Habitability Materials List" (Revision D) issued by the Naval Sea Systems Command (letter SEC 6101B/FJB 9640/0472H Ser. 1093). (See Appendix B for the Habitability List).

The U.S. Navy has conducted numerous investigations on proposed polymer usage in naval vessels, (e.g., fiberglass pipe as a substitute for metal pipe). This study considered such properties as fire resistance, joint quality, fatigue, etc. Additional information pertaining to uses of polymeric materials including relative cost, data on metals and polymers, appears in a study performed for the "National Shipbuilding Research Program on Plastics in Shipbuilding." (Maritime Administratition, 1977).

7.1.3 Coast Guard Vessels

Regulations pertaining to the use of polymeric materials on Coast Guard vessels are essentially those used for commercial vessels except where the naval specifications are more severe or are required for specific mission capability (see Section 7.5.1 and Table 7.4-7).

7.1.4 Fire Statistics

7.1.4.1 U.S. Coast Guard Regulated Vessels

A statistical summary of commercial vessel casualties due to fire and explosions, including deaths and injuries, appears in Tables 7.1-1 through 7.1.4 (U.S. Coast Guard, 1977, 1978). Tables 7.1-1 and 7.1-3 deal with the statistical summary of casualties to commercial vessels and Tables 7.1-2 and 7.1-4 deal with the statistical summary of deaths/injuries due to a vessel casualty for fiscal 1976 and 1977, respectively. Additional statistics concerning recreation and pleasure boating accidents are published in "Boating Statistics," CG-129, March 1977 and March 1978. (See Chapter 10).

Table 7.1.1. Statistical Summary of Casualties to Commercial Vessels[1]

1 July 1975 to 30 Sept. 1976 Fiscal year 1976[2]	Nature of Casualty																	
	Collisions: crossing, meeting and overtaking	Collisons, while anchored, docking or undocking	Collision, fog	Collisions with piers and bridges	Collisions, all others	Explosion and/or fires—cargo	Explosion and/or fires/vessel's fuel	Explosion and/or fire—boilers, pressure vessel	Explosion and/or fire—structure, equipment, all others	Grounding with damage	Grounding without damage	Founderings, capsizings and floodings	Heavy weather damage	Cargo damage	Material failure—structure and equipment	Material failure—machinery and engineering equipment	Casualty not otherwise classified	Total
Number of casualties	228	196	22	619	474	9	53	5	215	481	616	584	78	17	160	268	170	4211
Number of vessels involved	720	566	54	1204	894	12	57	5	219	823	1016	685	104	21	171	272	327	7150
Number of inspected vessels involved	177	136	15	388	297	7	10	1	61	202	300	66	63	12	125	165	107	2132
Number of uninspected vessels involved	543	430	39	816	597	5	47	4	158	621	716	619	41	9	46	107	220	5018
PRIMARY CAUSE																		
Personnel fault:																		
Pilots—State	8	8	0	43	6	0	0	0	0	7	31	0	0	0	0	2	5	113
Pilots—Federal	0	7	0	15	0	0	0	0	0	3	7	0	0	0	0	0	0	32
Licensed officer—documented seaman	131	75	14	355	170	1	2	0	5	174	185	31	2	0	4	4	17	1170
Unlicensed—undocumented persons	59	25	5	18	61	0	3	0	7	96	95	49	0	0	0	1	11	428
All others	16	13	1	46	21	4	4	0	13	16	27	12	0	6	9	1	29	218
Calculated risk	0	0	0	3	10	0	1	0	1	4	1	5	2	0	0	0	0	27
Restricted maneuvering room	12	21	4	44	27	0	0	0	6	25	35	19	1	2	1	3	10	210
Storms—adverse weather	2	22	3	31	49	0	1	0	0	52	53	66	89	10	18	1	28	405
Unusual currents	0	1	0	2	4	0	0	0	0	0	1	1	0	0	0	0	1	10
Sheer, suction, bank cushion	16	2	0	8	3	0	0	0	0	3	6	0	0	0	0	0	3	41
Depth of water, less than expected	0	2	0	4	10	0	0	0	0	88	268	1	0	0	0	0	0	373
Failure of equipment	24	43	1	65	54	0	36	5	63	61	70	67	1	1	68	220	97	879
Unseaworthy—lack of maintenance	0	0	0	0	1	0	0	0	0	0	0	30	0	0	3	0	0	34
Floating debris—submerged object	0	0	0	2	109	0	0	0	1	11	2	41	0	0	1	1	0	168
Inadequate tug assistance	1	1	0	16	4	0	0	0	0	3	3	0	0	0	0	0	0	28
Fault on part of other vessel or person	439	332	24	540	328	2	1	0	8	264	235	173	5	1	43	7	81	2483
Unknown—insufficient information	12	14	2	14	37	5	9	0	112	16	14	190	4	1	24	32	45	531

(Continued)

Table 7.1.1. Statistical Summary of Casualties to Commercial Vessels[1] (Continued)

1 July 1975 to 30 Sept. 1976 Fiscal year 1976[2]	Nature of Casualty																	
	Collisions: crossing, meeting and overtaking	Collisons, while anchored, docking or undocking	Collision, fog	Collisions with piers and bridges	Collisions, all others	Explosion and/or fires—cargo	Explosion and/or fires/vessel's fuel	Explosion and/or fire—boilers, pressure vessel	Explosion and/or fire—structure, equipment, all others	Grounding with damage	Grounding without damage	Founderings, capsizings and floodings	Heavy weather damage	Cargo damage	Material failure—structure and equipment	Material failure—machinery and engineering equipment	Casualty not otherwise classified	Total
TYPE OF VESSEL																		
Inspected vessels:																		
Passenger and ferry—large	4	3	0	14	7	0	0	0	1	5	3	0	0	0	4	7	4	52
Passenger and ferry—small	17	7	3	8	32	0	1	0	15	22	31	28	2	0	17	33	7	223
Freight	17	34	3	95	60	0	5	1	24	31	73	4	32	12	59	65	37	552
Cargo barge	13	9	0	16	22	1	0	0	2	21	16	10	13	0	6	0	7	136
Tankships	12	21	2	29	32	0	1	0	9	13	37	5	9	0	22	52	13	257
Tank barge	107	51	6	214	116	5	2	0	6	99	134	12	4	0	6	2	32	796
Public	1	3	0	0	2	0	0	0	1	2	2	1	1	0	1	4	1	19
Miscellaneous	6	8	1	12	26	1	1	0	3	9	4	6	2	0	10	2	6	97
Uninspected vessels:																		
Fishing	68	53	13	24	104	0	21	0	69	143	138	259	7	0	20	76	17	1012
Tugs	236	146	11	470	265	3	12	1	37	234	233	157	18	2	9	20	41	1895
Foreign	41	56	7	81	51	1	6	3	18	32	107	9	1	4	2	2	22	443
Miscellaneous	198	175	8	241	177	1	8	0	34	212	238	194	15	3	15	9	140	1668
GROSS TONNAGE																		
300 tons or less	376	318	34	478	504	4	44	2	142	304	423	562	37	5	65	134	125	3647
Over 300 to 1,000 tons	177	102	9	303	130	2	2	0	14	245	211	85	10	3	12	14	89	1410
Over 1,000 to 10,000 tons	127	80	7	272	177	6	4	0	31	132	197	24	23	2	40	34	59	1215
Over 10,000 tons	40	66	4	149	83	0	7	3	32	52	185	14	34	11	54	90	54	878
LENGTH																		
Less than 100 feet	316	258	32	373	415	4	35	1	124	334	318	504	28	3	55	123	91	3044
100 to less than 300 feet	338	209	13	616	338	7	13	2	43	405	446	166	30	6	33	25	154	2844

(Continued)

Table 7.1.1 Statistical Summary of Casualties to Commercial Vessels[1] (Continued)

1 July 1975 to 30 Sept. 1976 Fiscal year 1976[2]	Nature of Casualty																	
	Collisions; crossing, meeting and overtaking	Collisons, while anchored, docking or undocking	Collision, fog	Collisions with piers and bridges	Collisions, all others	Explosion and/or fires—cargo	Explosion and/or fires/vessel's fuel	Explosion and/or fire—boilers, pressure vessel	Explosion and/or fire—structure, equipment, all others	Grounding with damage	Grounding without damage	Founderings, capsizings and floodings	Heavy weather damage	Cargo damage	Material failure—structure and equipment	Material failure—machinery and engineering equipment	Casualty not otherwise classified	Total
300 to less than 500 feet	19	28	4	42	51	1	2	0	16	28	40	5	4	1	10	15	20	286
500 feet and over	47	71	5	173	90	0	7	2	36	56	182	10	42	11	73	109	62	976
AGE																		
Less than 10 years	381	220	27	577	393	9	27	3	83	394	489	182	61	10	82	116	149	3203
10 to less than 10 years	158	138	8	303	194	2	13	0	39	201	265	159	19	1	35	44	92	1674
20 to less than 30 years	70	59	5	139	101	1	6	1	34	95	103	123	11	1	14	27	30	823
30 years and over	111	149	14	182	206	0	11	1	63	133	156	221	13	9	40	85	56	1450
LOCATION OF CASUALTY																		
Inland—Atlantic	28	37	6	124	97	2	11	2	46	109	164	109	5	0	23	43	33	839
Inland—Gulf	100	66	7	180	102	3	16	2	48	72	176	92	4	3	8	13	44	936
Inalnd—Pacific	12	21	0	55	63	0	8	0	36	108	110	63	4	2	26	45	16	569
Ocean—Atlantic	9	6	6	1	23	0	4	0	17	13	16	54	20	2	20	58	12	261
Ocean—Gulf	15	9	0	4	51	0	8	0	12	23	9	65	5	1	13	20	7	242
Ocean—Pacific	12	2	0	0	40	0	5	0	23	28	11	80	30	8	30	52	6	327
Great Lakes	4	3	0	67	26	1	0	0	2	22	27	7	2	0	12	22	11	206
Western rivers	41	36	3	148	43	3	4	0	19	91	93	86	0	0	4	2	22	595
Ocean—other	0	2	0	2	3	0	0	0	6	5	3	3	3	0	4	6	4	41
Foreign waters	7	14	0	38	26	0	0	1	6	13	37	5	5	1	20	7	15	195
TIME OF DAY																		
Daylight	111	113	15	330	251	5	32	3	124	213	308	267	37	7	96	170	86	2168
Nighttime	103	64	7	258	176	3	17	0	81	235	287	225	23	9	39	78	56	1661
Twilight	14	19	0	31	47	1	7	2	10	36	51	72	18	1	25	20	28	382

(Continued)

Table 7.1.1 Statistical Summary of Casualties to Commercial Vessels[1] (Continued)

1 July 1975 to 30 Sept. 1976 Fiscal year 1976[2]	Nature of Casualty																	
	Collisions; crossing, meeting and overtaking	Collisons, while anchored, docking or undocking	Collision, fog	Collisions with piers and bridges	Collisions, all others	Explosion and/or fires—cargo	Explosion and/or fires/Vessel's fuel	Explosion and/or fire—boilers, pressure vessel	Explosion and/or fire—structure, equipment, all others	Grounding with damage	Grounding without damage	Founderings, capsizings and floodings	Heavy weather damage	Cargo damage	Material failure—structure and equipment	Material failure—machinery and engineering equipment	Casualty not otherwise classified	Total
ESTIMATED LOSSES (in thousands)																		
Vessel	9876	5092	582	8505	19092	1135	6129	300	22151	25660	557	34015	10793	781	1012	7410	7425	169812
Cargo	1214	116	10	6274	9160	212	33	0	2081	7027	5	3743	1956	379	689	7	232	33138
Property	65	1259	0	17958	6909	717	30	0	4491	1381	131	1686	749	23	1136	4	1202	37741
VESSELS TOTALLY LOST																		
Inspected	3	0	0	4	2	2	1	0	7	2	0	11	1	0	3	1	1	38
Uninspected	18	3	4	8	20	0	18	0	52	42	0	226	3	0	2	1	11	408

[1] Statistics concerning recreation and pleasure boating accidents are published in CG-357.
[2] Includes FY 76 Transition Quarter.

(Concluded)

Table 7.1.2 Statistical Summary of Deaths/Injuries Due to a Vessel Casualty[1]

1 July 1975 to 30 Sept. 1976 Fiscal Year 1976[2]	Nature of Casualty																	
	Collisions; crossing, meeting and overtaking	Collisons, while anchored, docking or undocking	Collision, fog	Collisions with piers and bridges	Collisions, all others	Explosion and/or fires—cargo	Explosion and/or fires/vessel's fuel	Explosion and/or fire—boilers, pressure vessel	Explosion and/or fire—structure, equipment, all others	Grounding with damage	Grounding without damage	Founderings, capsizings and floodings	Heavy weather damage	Cargo damage	Material failure—structure and equipment	Material failure—machinery and engineering equipment	Casualty not otherwise classified	Total
Number of casualties	15	6	3	6	7	4	8	4	24	2	7	72	1	1	9	4	13	186
Number of inspected vessels involved	1	2	0	2	0	3	3	0	4	1	1	1	0	1	4	2	7	32
Number of uninspected vessels involved	15	4	3	4	7	2	5	4	20	1	6	72	1	0	6	2	8	160
Number of persons deceased/injured	13/13	6/13	2/3	1/11	13/5	7/8	2/8	5/2	15/28	1/5	3/5	137/25	/1	1/	34/8	2/3	27/15	269/153
PRIMARY CAUSE																		
Personnel fault:																		
Pilots—State	—	—	—	—	—	—	—	—	—	—	—	—	—	—	—	—	—	0
Pilots—Federal	—	—	—	—	—	—	—	—	—	—	—	—	—	—	—	—	—	0
Licensed officer—documented seaman	1	—	—	2	—	1	2	—	1	1	—	3	—	—	—	—	1	12
Unlicensed—undocumented persons	4	3	2	—	1	—	—	—	3	—	2	12	—	—	—	—	—	27
All others	—	—	1	—	1	1	1	—	3	—	—	—	—	—	—	—	—	7
Error in judgement—calculated risk	—	—	—	—	—	—	—	—	—	—	—	—	—	—	—	—	—	0
Restricted manuevering room	2	—	—	—	1	—	—	—	—	—	—	3	—	—	—	—	1	7
Storms—adverse weather	—	—	—	2	—	—	1	—	—	—	1	12	1	—	1	—	1	19
Unusual currents	—	—	—	—	—	—	—	—	—	—	—	1	—	—	—	—	—	1
Sheer, suction, bank cushion	—	—	—	—	—	—	—	—	—	—	—	—	—	—	—	—	—	0
Depth of water less than expected	—	—	—	—	—	—	—	—	—	—	2	—	—	—	—	—	—	2
Failure of equipment	2	—	—	—	—	—	3	4	6	1	—	5	—	—	5	4	2	32
Unseaworthy—lack of maintenance	—	—	—	—	—	—	—	—	—	—	—	1	—	—	—	—	—	1
Floating debris—submerged object	—	—	—	—	—	—	—	—	—	—	—	1	—	—	—	—	—	1
Inadequate tug assistance	—	—	—	—	—	—	—	—	—	—	—	—	—	—	—	—	—	0
Fault on part of other vessel or person	7	3	—	1	3	1	—	—	1	—	—	2	—	—	2	—	4	24
Unknown—insufficient information	—	—	—	1	1	2	1	—	10	—	2	33	—	1	2	—	6	59

(Continued)

Table 7.1.2. Statistical Summary of Deaths/Injuries Due to a Vessel Casualty[1] (Continued)

1 July 1975 to 30 Sept. 1976 Fiscal Year 1976[2]	Nature of Casualty																	Total
	Collisions; crossing, meeting and overtaking	Collisions, while anchored, docking or undocking	Collision, fog	Collisions with piers and bridges	Collisions, all others	Explosion and/or fires—cargo	Explosion and/or fires/vessel's fuel	Explosion and/or fire—boilers, pressure vessel	Explosion and/or fire—structure, equipment, all others	Grounding with damage	Grounding without damage	Founderings, capsizings and floodings	Heavy weather damage	Cargo damage	Material failure—structure and equipment	Material failure—machinery and engineering equipment	Casualty not otherwise classified	
TYPE OF VESSEL INVOLVED																		
Inspected vessels:																		
Passenger and ferry—large	—	—	—	/4	—	—	—	—	—	—	—	—	—	—	—	—	/6	0/10
Passenger and ferry—small	/1	/9	—	—	—	—	—	—	—	/4	/2	—	—	—	—	—	/1	0/17
Freight	—	—	—	—	—	2/	—	4/2	—	—	—	1/	—	1/	29/4	/2	1/2	38/10
Cargo barge	—	—	—	—	—	—	—	—	—	—	—	—	—	—	—	—	—	0/0
Tankships	—	—	—	—	—	—	—	—	—	—	—	—	—	—	—	—	/1	0/1
Tank barges	—	/1	—	—	—	4/7	/1	—	/2	—	—	—	—	—	—	—	1/1	5/12
Public	—	—	—	—	—	—	—	—	—	—	—	—	—	—	—	—	—	0/0
Miscellaneous	—	—	—	1/2	—	—	/3	—	—	—	—	5/	—	—	1/1	/1	3/	10/7
Uninspected vessels:																		
Fishing	1/4	—	/1	—	4/3	—	/2	—	7/7	1/1	3/2	66/9	/1	—	1/	2/	1/1	86/31
Tugs	1/3	—	—	/5	—	3/	/2	1/	1/3	—	/1	13/7	—	—	—	—	1/3	20/24
Foreign	—	—	2/	—	7/	—	—	4/2	2/4	—	—	1/	—	—	1/	—	17/	34/6
Miscellaneous	11/5	6/3	/2	—	2/2	/1	—	—	1/10	—	—	51/9	—	—	2/3	—	3/	76/35
PARTICULARS OF PERSON DECEASED/INJURED																		
Papers of deceased/injured:																		
Licensed by Coast Guard	1/1	—	—	/2	—	/1	/3	—	/2	/1	—	6/3	—	—	8/1	—	/1	15/15
Documented by Coast Guard	/1	—	—	/2	—	—	/1	—	/1	—	—	3/1	—	—	21/1	—	1/2	25/9
No license or document	12/10	6/13	/3	1/7	13/5	7/7	/4	3/1	13/23	1/4	3/5	126/21	/1	1/	4/6	2/3	9/12	201/125
Other—unknown—foreign	/1	—	2/	—	—	—	2/	2/1	2/2	—	—	2/	—	—	1/	—	17/	28/4
Status or capacity on vessel:																		
Passenger	3/1	/9	/1	—	2/	—	—	—	1/6	/1	1/3	31/1	—	—	—	—	2/6	40/28

(Continued)

Table 7.1.2. Statistical Summary of Deaths/Injuries Due to a Vessel Casualty[1] (Continued)

1 July 1975 to 30 Sept. 1976 Fiscal Year 1976[2]	Collisions; crossing, meeting and overtaking	Collisions, while anchored, docking or undocking	Collision, fog	Collisions with piers and bridges	Collisions, all others	Explosion and/or fires—cargo	Explosion and/or fires/vessel's fuel	Explosion and/or fire—boilers, pressure vessel	Explosion and/or fire—structure, equipment, all others	Grounding with damage	Grounding without damage	Founderings, capsizings and floodings	Heavy weather damage	Cargo damage	Material failure—structure and equipment	Material failure—machinery and engineering equipment	Casualty not otherwise classified	Total
Longshoreman—harbor worker	—	—	—	—	—	—	—	—	3/	—	—	—	—	1/	/4	—	/2	4/6
Crewmember	8/9	6/4	2/2	1/11	11/5	1/6	/8	5/2	10/18	1/4	2/2	84/19	/1	—	32/3	2/3	25/6	190/103
Other	2/3	—	—	—	—	6/2	2/	—	1/4	—	—	22/5	—	—	2/1	—	/1	35/16
Activity engaged in:																		
Off duty	3/4	—	—	1/2	7/	/1	--	—	1/5	—	—	5/5	—	—	22/	—	—	39/17
Deck department duties	2/3	1/4	/2	/5	3/4	1/2	/1	—	3/4	/3	/2	15/7	/1	—	6/3	—	7/4	38/45
Engine department duties	/1	—	—	/2	—	/3	/3	5/2	1/7	—	—	1/2	—	—	4/	/3	1/1	12/24
Stewards department duties	—	—	—	/1	—	—	—	—	2/1	/1	/1	1/	—	—	—	—	—	3/4
Handling cargo	—	—	—	—	—	—	—	—	—	—	—	—	—	1/	1/5	—	/2	2/7
Fishing	3/3	—	—	—	1/1	—	/3	—	1/1	1/	3/	44/4	—	—	—	2/	1/1	56/13
Drills	—	/9	/1	—	—	—	—	—	—	—	—	—	—	—	—	—	—	/10
Passenger	3/	—	—	—	2/	—	—	—	1/5	/1	/2	19/2	—	—	—	—	1/6	26/16
Other and unknown	2/2	5/	2/	/1	—	6/2	2/1	—	6/5	—	—	52/5	—	—	1/	—	17/1	93/17
Location of vessel:																		
At dock	—	—	/1	—	—	5/2	2/	—	/2	—	—	/1	—	—	—	—	—	7/6
At anchor	6/3	1/9	—	1/1	/3	/1	/1	—	/3	—	—	9/3	—	—	29/	—	2/	48/24
Underway	7/10	5/4	2/2	/10	13/2	2/5	/7	5/2	15/28	1/5	3/5	128/21	/1	1/	5/8	2/3	25/15	214/123
Unknown	—	—	—	—	—	—	—	—	—	—	—	—	—	—	—	—	—	0/0
PART OF BODY INVOLVED																		
Head and upper limbs	/3	—	/2	/4	—	1/2	—	1/	/4	/1	—	/7	—	—	2/1	/1	2/4	6/29
Back and lower limbs	—	—	—	—	/1	—	/1	—	/5	/1	—	/3	—	—	—	—	/2	/13
Chest	/1	/1	—	/1	—	—	/1	—	1/	—	—	/5	/1	—	/1	—	—	1/11
Extremities	/2	/2	—	/1	/2	/1	/3	—	/16	/2	/5	/4	—	—	/2	/1	/4	/45

(Continued)

114

Table 7.1.2 Statistical Summary of Deaths/Injuries Due to a Vessel Casualty[1] (Continued)

1 July 1975 to 30 Sept. 1976 Fiscal Year 1976[2]	Nature of Casualty																	
	Collisions; crossing, meeting and overtaking	Collisions, while anchored, docking or undocking	Collision, fog	Collisions with piers and bridges	Collisions, all others	Explosion and/or fires—cargo	Explosion and/or fires/vessel's fuel	Explosion and/or fire—boilers, pressure vessel	Explosion and/or fire—structure, equipment, all others	Grounding with damage	Grounding without damage	Founderings, capsizings and floodings	Heavy weather damage	Cargo damage	Material failure—structure and equipment	Material failure—machinery and engineering equipment	Casualty not otherwise classified	Total
Illness	—	—	/1	—	—	2/	—	—	—	—	—	1/	—	—	—	—	—	3/1
Drowning	11/8	6/	—	—	3/	—	—	—	—	1/	2/	75/	—	—	1/1	1/	3/	103/1
Miscellaneous and unspecified	2/7	/10	2/	1/5	10/2	4/5	2/3	4/2	14/3	/1	1/	61/6	—	1/	31/3	1/1	22/5	156/53

[1] Statistics concerning recreation and pleasure boating accidents are published in CG-357.
[2] Includes FY76 Transition Quarter.

Table 7.1.3 Statistical Summary of Casualties to Commercial Vessels

1 October 1976 to 30 September 1977 Fiscal Year 1977	Collisions; crossing, meeting and overtaking	Collisons, while anchored, docking or undocking	Collision, fog	Collisions with piers and bridges	Collisions, all others	Explosion and/or fires—cargo	Explosion and/or fires/vessel's fuel	Explosion and/or fire—boilers, pressure vessel	Explosion and/or fire—structure, equipment, all others	Grounding with damage	Grounding without damage	Founderings, capsizings and floodings	Heavy weather damage	Cargo damage	Material failure—structure and equipment	Material failure—machinery and engineering equipment	Casualty not otherwise classified	Total
Number of casualties	269	218	3	493	280	16	8	9	181	417	650	414	23	24	213	286	70	3574
Number of vessels involved	830	568	9	958	485	20	10	9	193	766	1077	498	34	26	243	308	106	6140
Number of inspected vessels involved	213	159	1	295	161	7	1	6	55	210	324	45	13	21	135	170	36	1852
Number of uninspected vessels involved	617	409	8	663	324	13	9	3	138	556	753	453	21	5	108	138	70	4288
PRIMARY CAUSE																		
Personnel fault:																		
Pilots—State	15	7	—	22	2	—	—	—	—	8	34	2	—	—	—	—	3	93
Pilots—Federal	2	4	—	15	—	—	—	—	—	2	14	—	—	—	—	—	—	37
Licensed officer—documented seaman	144	107	1	274	89	2	—	—	3	137	207	30	2	1	10	2	10	1019
Unlicensed—undocumented persons	71	29	1	20	17	—	1	1	8	70	84	37	3	—	6	1	10	359
All others	12	20	—	33	8	8	—	1	9	9	37	8	—	5	6	3	5	164
Calculated risk	—	—	—	—	2	—	—	—	—	2	3	1	—	—	—	—	—	8
Restricted maneuvering room	2	3	—	9	1	—	—	—	—	3	3	—	—	—	—	1	—	22
Storms—adverse weather	8	10	—	14	30	—	—	—	1	31	38	64	18	17	53	4	10	298
Unusual currents	3	2	—	6	2	—	—	—	—	7	3	3	—	—	—	—	4	30
Sheer, suction, bank cushion	11	1	—	4	1	—	—	—	—	6	6	—	—	—	—	—	2	31
Depth of water less than expected	1	2	—	3	6	—	—	—	—	64	155	3	—	—	—	—	—	234
Failure of equipment	14	21	—	54	14	5	1	4	67	35	46	63	1	1	70	256	7	659
Unseaworthy—lack of maintenance	—	—	—	2	1	—	—	—	1	1	2	63	—	—	14	5	—	89
Floating debris—submerged object	—	—	—	13	90	—	—	—	—	11	1	20	—	—	6	—	—	141
Inadequate tug assistance	1	1	—	6	1	—	—	—	—	—	4	—	—	—	—	—	—	13
Fault on part of other vessel or person	542	351	7	475	208	4	3	—	13	355	429	141	10	2	65	23	36	2664
Unknown—insufficient information	4	10	—	8	13	1	5	3	91	25	11	63	—	—	13	13	19	279
TYPE OF VESSEL																		
Inspected vessels:																		
Passenger and ferry—large	3	1	—	12	2	—	—	—	1	—	4	—	—	2	5	8	1	39

(Continued)

Table 7.1.3 Statistical Summary of Casualties to Commercial Vessels (Continued)

1 October 1976 to 30 September 1977 Fiscal Year 1977	Collisions; crossing, meeting and overtaking	Collisons, while anchored, docking or undocking	Collision, fog	Collisions with piers and bridges	Collisions, all others	Explosion and/or fires—cargo	Explosion and/or fires/vessel's fuel	Explosion and/or fire—boilers, pressure vessel	Explosion and/or fire—structure, equipment, all others	Grounding with damage	Grounding without damage	Founderings, capsizings and floodings	Heavy weather damage	Cargo damage	Material failure—structure and equipment	Material failure—machinery and engineering equipment	Casualty not otherwise classified	Total
Passenger and ferry—small	10	7	—	6	8	—	—	—	14	20	14	21	1	—	6	20	3	130
Freight	26	47	—	79	34	—	—	4	26	21	68	6	5	19	60	82	10	487
Cargo barge	2	—	—	4	5	—	—	—	—	1	1	—	—	—	2	—	1	16
Tankships	9	15	—	25	19	1	—	2	4	16	42	6	4	—	31	37	3	214
Tank barge	154	70	1	167	87	5	—	—	8	147	191	7	3	—	20	14	15	889
Public	6	11	—	1	2	—	—	—	2	3	4	5	—	—	3	6	2	45
Miscellaneous	3	8	—	1	4	1	1	—	—	2	—	—	—	—	8	3	1	32
Uninspected vessels:																		
Fishing	98	60	2	22	34	1	4	1	61	93	125	203	4	—	33	85	17	843
Tugs	255	139	3	355	174	3	2	—	37	235	310	98	10	3	30	28	24	1706
Foreign	53	79	1	54	19	6	—	1	13	27	101	4	1	—	6	12	14	391
Miscellaneous	211	131	2	232	97	3	3	1	27	201	217	148	6	2	39	13	15	1348
GROSS TONNAGE																		
300 tons or less	437	287	4	374	232	6	9	2	132	326	402	405	16	6	90	141	56	2925
Over 300 to 1,000 tons	167	103	1	248	87	4	1	—	15	240	290	66	4	2	24	9	8	1270
Over 1,000 to 10,000 tons	174	102	3	234	121	7	—	3	22	159	222	15	8	6	61	52	23	1212
Over 10,000 tons	52	76	—	102	45	3	—	4	24	41	163	12	6	12	68	106	19	733
LENGTH																		
Less than 100 feet	374	235	4	305	178	3	9	1	118	274	323	356	10	3	69	124	47	2433
100 to less than 300 feet	380	221	3	503	221	11	1	1	36	423	535	128	13	3	78	47	24	2628
300 to less than 500 feet	26	21	2	26	38	2	—	2	13	20	45	3	4	6	17	17	10	252
500 feet and over	50	91	—	124	48	4	—	5	26	49	174	11	7	14	79	120	25	827
AGE																		
Less than 10 years	395	241	5	434	194	10	5	2	72	326	481	119	12	9	77	118	47	2547
10 to less than 20 years	194	130	3	233	123	5	1	3	49	210	278	111	11	6	57	62	24	1500
20 to less than 30 years	77	67	1	132	68	1	1	1	23	106	138	88	5	2	47	51	10	818
30 years and over	164	130	—	159	100	4	3	3	49	124	180	180	6	9	62	77	25	1275
LOCATION OF CASUALTY																		
Inland—Atlantic	36	39	—	90	65	4	—	2	34	84	210	85	2	1	40	64	18	774

(Continued)

Table 7.1.3. Statistical Summary of Casualties to Commercial Vessels (Continued)

1 October 1976 to 30 September 1977 Fiscal Year 1977	Collisions; crossing, meeting and overtaking	Collisions, while anchored, docking or undocking	Collision, fog	Collisions with piers and bridges	Collisions, all others	Explosion and/or fires—cargo	Explosion and/or fires/vessel's fuel	Explosion and/or fire—boilers, pressure vessel	Explosion and/or fire—structure, equipment, all others	Grounding with damage	Grounding without damage	Founderings, capsizings and floodings	Heavy weather damage	Cargo damage	Material failure—structure and equipment	Material failure—machinery and engineering equipment	Casualty not otherwise classified	Total
Inland—Gulf	108	55	1	137	63	5	3	2	45	80	112	74	4	1	18	15	11	734
Inland—Pacific	22	28	—	46	33	2	1	—	29	64	79	67	2	2	23	30	5	433
Ocean—Atlantic	9	4	—	5	13	1	—	1	11	14	6	47	4	8	39	55	13	230
Ocean—Gulf	12	13	—	15	16	1	—	—	14	11	6	25	2	1	10	21	2	149
Ocean—Pacific	15	7	—	3	8	—	2	2	19	21	10	38	5	6	24	40	4	204
Great Lakes	6	9	—	50	26	—	—	1	7	22	42	9	1	—	19	35	5	232
Western rivers	53	36	2	116	47	3	2	—	15	109	154	66	—	1	17	7	7	635
Ocean—other	3	12	—	13	5	—	—	1	5	5	14	2	2	3	15	10	2	92
Foreign waters	5	15	—	18	4	—	—	—	2	7	17	1	1	1	8	9	3	91
TIME OF DAY																		
Daylight	126	126	1	254	141	12	6	4	105	175	311	210	10	16	94	176	38	1805
Nighttime	126	84	2	189	99	4	—	3	65	186	297	160	9	6	63	84	22	1399
Twilight	17	8	—	50	40	—	2	2	11	56	42	44	4	2	56	26	10	370
ESTIMATED LOSSES ($1000's)																		
Vessel	10459	5774	35	6804	10945	3842	677	492	26762	36335	0	21119	993	189	5440	7348	1927	139141
Cargo	678	19	0	610	1211	18	2	1	1237	2552	0	3957	91	4612	475	1007	719	17189
Property	2137	417	53	16577	640	3	5	0	6486	6754	737	683	0	26	58	84	227	34887
VESSELS TOTALLY LOST																		
Inspected	—	—	—	—	2	3	—	—	3	2	—	9	1	—	2	—	—	22
Uninspected	20	3	—	8	18	3	3	2	39	35	—	123	1	—	14	1	2	272

(Concluded)

Table 7.1.4 Statistical Summary of Deaths/Injuries Due to a Vessel Casualty

1 October 1976 to 30 September 1977 Fiscal Year 1977	Collisions; crossing, meeting and overtaking	Collisions, while anchored, docking or undocking	Collision, fog	Collisions with piers and bridges	Collisions, all others	Explosion and/or fires—cargo	Explosion and/or fires/vessel's fuel	Explosion and/or fire—boilers, pressure vessel	Explosion and/or fire—structure, equipment, all others	Grounding with damage	Grounding without damage	Founderings, capsizings and floodings	Heavy weather damage	Cargo damage	Material failure—structure and equipment	Material failure—machinery and engineering equipment	Casualty not otherwise classified	Total
Number of casualties	21	8	1	5	2	9	4	2	19	4	1	41	0	0	6	5	5	133
Number of inspected vessels involved	1	2	0	1	0	4	0	1	7	1	0	5	0	0	3	2	1	28
Number of uninspected vessels involved	20	6	1	4	2	5	4	1	12	3	1	36	0	0	3	3	4	105
Number of persons deceased/injured	88/34	1/17	0/1	1/9	3/1	19/16	0/5	0/5	14/19	1/3	0/1	74/15	0/0	0/0	8/3	3/4	4/3	216/136
PRIMARY CAUSE																		
Personnel fault:																		
Pilots—State	—	—	—	—	—	—	—	—	—	—	—	—	—	—	—	—	—	0
Pilots—Federal	—	—	—	—	—	—	—	—	—	—	—	—	—	—	—	—	—	0
Licensed officer—documented seaman	3	1	—	3	1	1	—	—	1	1	—	4	—	—	—	—	1	16
Unlicensed—undocumented persons	8	1	—	1	—	—	—	—	5	2	—	4	—	—	—	—	—	21
All others	—	—	—	—	—	3	—	1	1	—	—	1	—	—	—	—	—	6
Error in judgement—calculated risk	—	—	—	—	—	—	—	—	—	—	—	—	—	—	—	—	—	0
Restricted maneuvering room	—	—	—	—	—	—	—	—	—	—	—	—	—	—	—	—	—	0
Storms—adverse weather	—	—	—	—	—	—	—	—	1	—	1	15	—	—	—	—	1	18
Unusual currents	—	—	—	—	—	—	—	—	—	—	—	1	—	—	—	—	—	1
Shear, suction, bank cushion	—	—	—	—	—	—	—	—	—	—	—	—	—	—	—	—	—	0
Depth of water less than expected	—	—	—	—	—	—	—	—	—	—	—	1	—	—	—	—	—	1
Failure of equipment	—	—	—	—	—	1	—	—	—	—	—	4	—	—	4	5	—	19
Unseaworthy—lack of maintenance	—	—	—	—	—	—	—	—	—	—	—	1	—	—	—	—	—	1
Floating debris—submerged object	—	—	—	—	1	—	—	—	—	—	—	1	—	—	—	—	—	2
Inadequate tug assistance	—	—	—	—	—	—	—	—	—	—	—	—	—	—	—	—	—	0
Fault of other vessel or person	10	6	1	1	—	2	—	—	—	—	—	4	—	—	1	—	1	26
Unknown—insufficient information	—	—	—	—	—	1	4	1	8	—	—	5	—	—	1	—	2	22
TYPE OF VESSEL INVOLVED																		
Inspected vessels:																		
Passenger and ferry—large	—	—	—	/1	—	—	—	—	/1	—	—	—	—	—	—	/1	—	0/3
Passenger and ferry—small	—	/1	—	—	—	—	—	—	—	/1	—	15/	—	—	—	—	—	15/2

(Continued)

Table 7.1.4 Statistical Summary of Deaths/Injuries Due to a Vessel Casualty (Continued)

1 October 1976 to 30 September 1977 Fiscal Year 1977	Collisions; crossing, meeting and overtaking	Collisions, while anchored, docking or undocking	Collision, fog	Collisions with piers and bridges	Collisions, all others	Explosion and/or fires—cargo	Explosion and/or fires/vessel's fuel	Explosion and/or fire—boilers, pressure vessel	Explosion and/or fire—structure, equipment, all others	Grounding with damage	Grounding without damage	Founderings, capsizings and floodings	Heavy weather damage	Cargo damage	Material failure—structure and equipment	Material failure—machinery and engineering equipment	Casualty not otherwise classified	Total
Freight	/1	—	—	—	—	—	—	—	2/8	—	—	1/	—	—	1/	—	—	4/9
Cargo barge	—	—	—	—	—	—	—	—	—	—	—	—	—	—	—	—	—	0/0
Tankships	—	—	—	—	—	—	—	/2	—	—	—	1/6	—	—	3/	—	—	4/8
Tank barges	—	/1	—	—	—	4/5	—	/1	1/	—	—	—	—	—	—	1/	—	5/8
Public	—	—	—	—	—	—	—	—	—	—	—	—	—	—	—	—	—	0/0
Miscellaneous	—	—	—	—	—	2/2	—	—	—	—	—	—	—	—	—	—	/1	2/3
Uninspected vessels:																		
Fishing	3/11	/2	—	—	3/	1/	/1	—	3/2	1/1	—	37/5	—	—	3/	—	2/1	53/23
Tugs	/1	/3	—	1/6	/1	—	/3	—	2/1	/1	/1	7/2	—	—	/3	—	—	10/22
Foreign	—	—	—	—	—	9/9	—	—	/2	—	—	—	—	—	—	/1	—	10/11
Miscellaneous	85/21	1/10	/1	/2	—	3/	/1	/2	6/5	—	—	13/2	—	—	1/	2/2	2/1	113/47
PARTICULARS OF PE RSON DECEASED/INJURED																		
Paper of deceased/injured:																		
Licensed by Coast Guard	—	—	—	—	—	—	—	—	—	—	—	—	—	—	—	—	/1	1/2
Documented by Coast Guard	/1	/1	—	/1	—	—	—	/2	2/6	/1	/1	2/7	—	—	3/	—	—	7/20
No license or document	88/33	1/15	/1	1/8	3/1	10/7	/5	/3	11/9	1/2	—	72/8	—	—	4/3	2/4	4/2	197/101
Other—unknown—foreign	—	/1	—	—	—	9/9	—	—	/3	—	—	—	—	—	1/	1/	—	11/13
Status or capacity on vessel:																		
Passenger	76/18	1/9	/1	/1	—	—	—	—	2/2	/1	—	18/1	—	—	—	/1	/1	97/35
Longshoreman—harbor worker	—	—	—	—	—	—	—	—	—	—	—	—	—	—	—	1/1	—	1/1
Crewmember	12/16	/8	—	/6	3/1	16/13	/5	/3	10/10	1/2	/1	56/14	—	—	7/3	2/1	4/2	111/85
Other	—	—	—	1/2	—	3/3	—	/2	2/7	—	—	—	—	—	1/	/1	—	7/15
Activity engaged in:																		
Off duty	/3	/1	—	/2	—	—	/1	—	3/1	—	—	1/	—	—	—	—	—	4/8
Deck department duties	11/9	/5	—	/2	—	12/7	/2	/2	1/5	/1	/1	16/3	—	—	3/3	1/1	4/1	48/42

(Continued)

Table 7.1.4 Statistical Summary of Deaths/Injuries Due to a Vessel Casualty (Continued)

1 October 1976 to 30 September 1977 Fiscal Year 1977	Collisions; crossing, meeting and overtaking	Collisions, while anchored, docking or undocking	Collision, fog	Collisions with piers and bridges	Collisions, all others	Explosion and/or fires—cargo	Explosion and/or fires/vessel's fuel	Explosion and/or fire—boilers, pressure vessel	Explosion and/or fire—structure, equipment, all others	Grounding with damage	Grounding without damage	Founderings, capsizings and floodings	Heavy weather damage	Cargo damage	Material failure—structure and equipment	Material failure—machinery and engineering equipment	Casualty not otherwise classified	Total
Engine department duties	/1	/1	—	/1	—	4/3	/2	—	1/1	—	—	—	—	—	—	—	—	5/9
Stewards department duties	/1	—	—	—	—	/3	—	—	—	—	—	2/	—	—	—	—	—	2/4
Handling cargo	—	—	—	—	—	—	—	—	/	—	—	/1	—	—	1/	1/1	—	2/3
Fishing	1/2	/2	—	—	3/	—	—	—	2/	1/1	—	36/4	—	—	3/	—	/1	46/10
Drills	—	—	—	—	—	—	—	—	—	—	—	—	—	—	—	—	—	0/0
Passenger	76/18	1/8	/1	/1	—	—	—	—	2/2	/1	—	4/1	—	—	—	/1	/1	83/34
Other and unknown	—	—	—	1/3	/1	3/3	—	/3	5/9	—	—	15/6	—	—	1/	1/1	—	26/26
Location of vessel:																		
At dock	—	/4	—	—	—	1/2	/1	—	1/1	—	—	—	—	—	1/	—	—	3/8
At anchor	10/18	/2	—	1/2	—	—	—	/1	—	—	—	3/	—	—	/3	—	—	14/26
Underway	78/16	1/11	/1	/7	3/1	18/14	/4	/4	13/18	1/3	/1	71/15	—	—	7/	3/4	4/3	199/102
Unknown	—	—	—	—	—	—	—	—	—	—	—	—	—	—	—	—	—	0/0
PART OF BODY INVOLVED																		
Head and upper limbs	/1	/6	—	/3	/1	/5	/2	—	/2	/1	—	—	—	—	—	—	3/1	3/22
Back and lower limbs	/1	/7	/1	/3	—	/1	/1	—	/2	—	—	/1	—	—	—	—	/2	0/19
Chest	/1	/1	—	/1	—	—	—	—	/1	/1	—	/1	—	—	—	—	—	0/6
Extremities	/4	/2	—	—	—	/5	/1	/5	1/4	/1	/1	/4	—	—	/3	/2	—	1/32
Illness	—	—	—	/1	—	—	—	—	—	—	—	1/1	—	—	—	—	1/	2/2
Drowning	85/	1/	—	1/	2/	1/	—	—	2/	—	—	53/	—	—	4/	—	1/	150/0
Miscellaneous and unspecified	3/27	s/1	—	/1	1/	18/5	/1	—	11/10	1/	—	20/8	—	—	4/	/1	2/	60/55

(Concluded)

It is interesting to note that the vessel casualty statistics, in Tables 7.1-1 and 7.1-3, relating to *explosions and fires* are significantly less than those for other types of casualties. (See heavy lined areas in the tables). Coast Guard-inspected vessels are safer than uninspected vessels. (Refer to Chapter 3, Section 3.2, for additional information pertaining to fire statistics).

A commercial vessel is a completely self-contained unit that transports people and cargo over prescribed distances with reasonable economy and virtually complete independence of assistance. Despite this independence, fires do occur. They can be classified as (1) pier or shipyard fires, (2) in-transit fires, and (3) post-collision fires.

(1) *Pier or Shipyard Fires.* This type of fire involves all types of surface vessels. Personnel aboard during these incidents could include ship's crew, passengers, cargo transfer personnel, or shipyard personnel. The vessel could be tied-up at pierside, at a loading or unloading facility, at a shipyard repair dock or some other repair facility, or could still be under construction. The causes of these fires vary greatly. The major areas of concern are the electrical system, tank cleaning/working operations, engine room, and the passenger and crew accommodation spaces. Hotwork (welding and burning) resulting from repairs or alterations is a major cause of fires under these conditions.

(2) *In-Transit Fires.* These fires usually occur in the engineering, machinery, and cargo spaces and spread to the passenger and crew accommodation areas. They frequently start as a result of electrical or mechanical malfunction or cigarette smoking. A classic example is the fire that occurred on the M/V Cunard Ambassador. (See Section 3.4.2).

(3) *Post-Collision Fires.* These fires usually occur after a collision between two ships; a moving ship striking a pier or bridge; or a moving ship striking one that is anchored. A classic example of this latter type is the collision between the SS C.V. Seawitch and the SS Esso Brussels (See Section 3.4.2).

In-transit and post-collision fires are potentially catastrophic but are usually controlled so that few fatalities occur. Some exceptions to this rule, however, such as the SS Morro Castle, SS Yarmouth Castle, SS Key Trader and SS Baune, SS C.V. Seawitch and SS Esso Brussels, and the M/V Cunard Ambassador are examples of catastrophes where many lives and considerable cargo were lost. (There were no fatalities aboard the M/V Cunard Ambassador). Fires following collisions especially those involving oil spills are a serious problem.

Figures 7.1-1 through 7.1-5 are representative of shipboard fire damage. The M/V Cunard Ambassador fire damage photos (Figures 7.1-6 through 7.1-25) show that while the damage was extensive ($10 million), the ship did not sink. The fire-resistant bulkheads contained or slowed the progress of the fire sufficiently to allow extinguishment. This demonstrated the success of shipbuilding regulation (in this case, Subchapter H-Passenger Vessels). (See Chapter 3, Sections 3.4.1 and 3.4.2 for additional information).

Figure 7.1.1. Representative shipboard fire damage.

Figure 7.1.2. Representative shipboard fire damage.

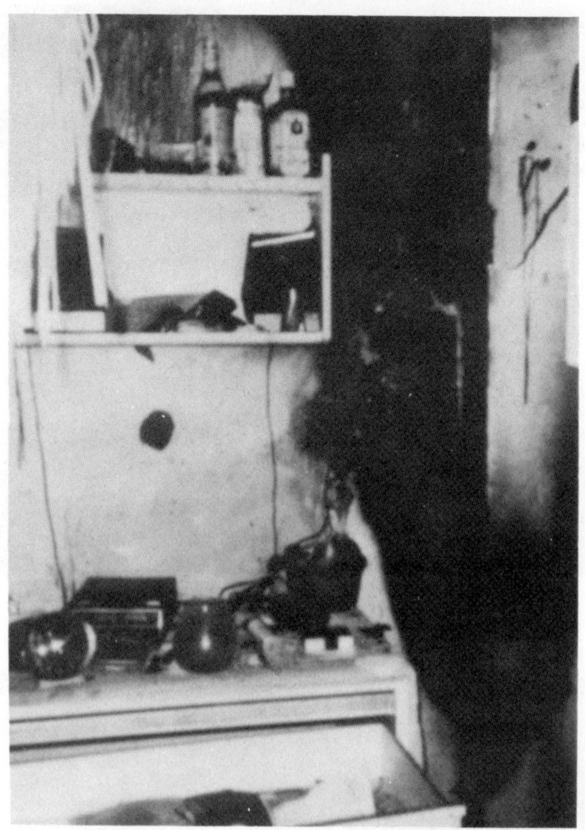

Figure 7.1.3. Representative shipboard fire damage.

The statistics in Tables 7.1-1 through 7.1-4 indicate, as said before in this volume, that while fires do occur on ships, resulting in injuries, deaths, and property loss or damage, the problem is not as severe as fire related injuries, death and property damage in land transportation vehicles, or buildings. Overall these statistics seem to indicate a good fire safety effort for Coast Guard-regulated vessels.

The statistical summaries presented in Tables 7.1-1 through 7.1-4 represent reportable casualties involving commercial vessels. It is important to note that the fiscal 1976 casualties (Tables 7.1-1 and 7.1-3) reported by Coast Guard Headquarters represent an extended fiscal year of 15 months (1 July, 1975, to 30 September, 1976) as compared to a 12-month period for fiscal 1977 (1 October, 1976 to 30 September 1977). This extended period should be taken into consideration in any comparison of statistics of previous fiscal years. Briefly summarized, the

124

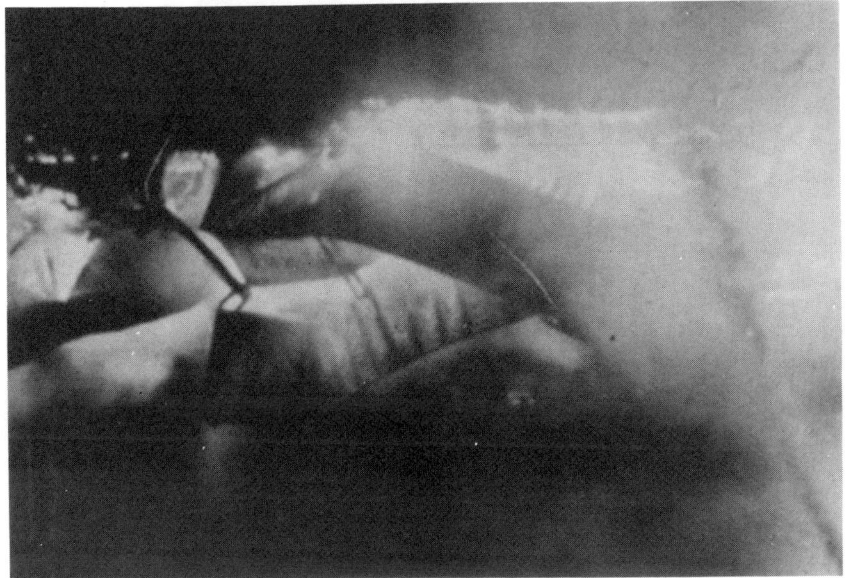

Figure 7.1.4. Representative shipboard fire damage.

major marine casualties for fiscal 1976 and 1977 are as follows: 46 deaths, 7 injuries, and $25,790,000 property damage for fiscal 1976; as compared to 87 deaths, 33 injuries, and $2,890,000 property damage for fiscal 1977.

7.1.4.2 Mobile Offshore Drilling Structures-Regulations and Statistics

Regulations pertaining to the use of polymeric materials on offshore drilling structures are essentially the same as those used for commercial vessels. (See Section 7.5.1 and Table 7.4-5). Table 7.1-5 reproduces statistics pertaining to mishaps involving mobile offshore drilling structures from 1955 to 1974. The information appeared in the "Offshore News" dated June 5, 1974. It is interesting to note that storms were the major cause of mishaps, and not blowouts and fires. (There were no incidents in which damage was initiated by fire in structural materials).

7.1.4.3 U.S. Navy Vessels

Fire has been a serious cause of damage aboard naval vessels, even in peacetime. Fire aboard aircraft carriers has been especially costly. Since 1965 the cost of fire damage in dollars as listed in Table 7.1-6 aboard approximately a dozen U.S. aircraft carriers has been in excess of $100 million (based upon initial cost information obtained from the Navy Safety Center).

Figure 7.1.5. Representative shipboard fire damage.

Figure 7.1.6. Fire origin starboard diesel.

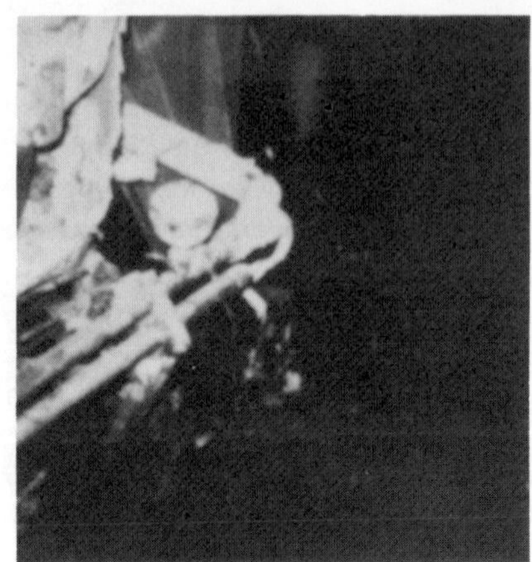

Figure 7.1.7. Port diesel.

Figure 7.1.8. Hydraulic control room.

Figure 7.1.9. Watertight door.

Figure 7.1.10. Generator room fire pump.

Figure 7.1.11. Damaged stateroom.

The Oriskany fire was caused by a Mark 24 flare. It resulted in 43 casualties and $10 million damage. In the case of the Enterprise, a flight deck fire caused secondary burning of aircraft fuel and ordnance, with extensive damage to the flight decks. Table 7.1-6 shows the 1967 estimated damage to the Forrestal, which was finally estimated to cost more than $80 million in repairs, exclusive of replacement of aircraft. The Forrestal fire burned for a day after penetrating the decks. In the case of the Saratoga, the fire hazard was the cable insulation; the fire spread along cableways from compartment to compartment, due to the flammability of the insulation around the cables.

Figure 7.1.12. Control room.

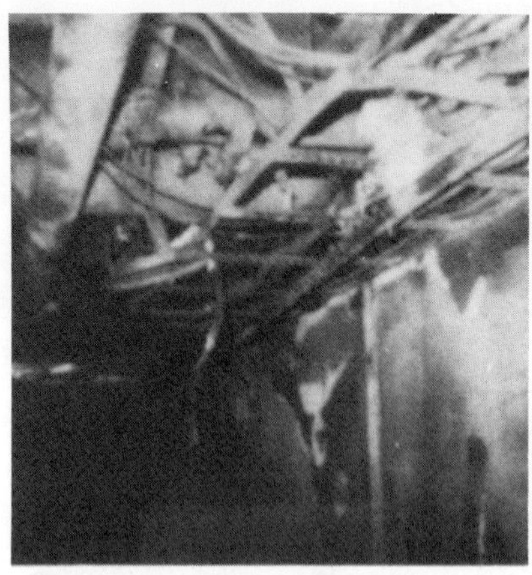

Figure 7.1.13. Corridor outside control room.

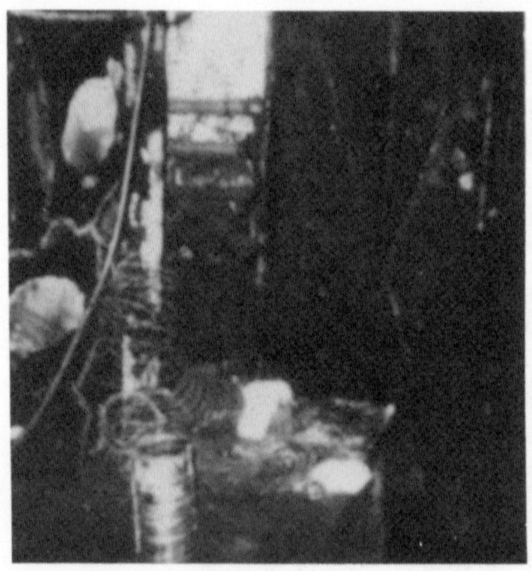

*Figure 7.1.14. Hydraulic control room
entrance.*

*Figure 7.1.15. Undamaged stateroom beside
hydraulic control room.*

Of particular interest for the purposes of this discussion is the second Forrestal fire (see 7.8.3.1.1), which was initiated by sabotage in the flag officer's quarters. Despite the controls on the compartment contents and specifications controlling materials used to limit flammability, the furnishings and decorations, most of which were polymers, were sufficiently flammable to propagate an intense fire, resulting in heavy damage to the vessel.

In addition to the carrier fires, several major fires have also occurred recently in other types of combatant ships, some of which are listed in Table 7.1-7. Although

Figure 7.1.16. Shop area.

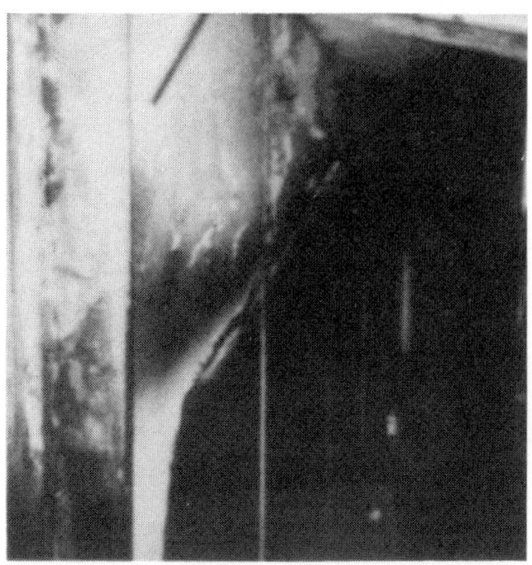

Figure 7.1.17. Stairwell.

Figure 7.1.18. Undamaged stateroom.

Figure 7.1.19. Lounge.

highly flammable ordnance and fuel were major contributors to the cause and propagation of these fires, in most cases polymers of various types were secondary fuel sources, adding to the intensity of the fires.

Fire is a special hazard for High Performance Ships, whose hulls and superstructure are constructed predominantly of aluminum because of the pressing need for weight reduction, especially above the water line. because the melting point of aluminum (640°C) is lower than the temperature attained in most major fires, serious structural distortion often results, thus raising significantly the damage costs. In some cases, the entire structure is destroyed when the aluminum melted. The subject will receive special attention in Chapter 8.

Figure 7.1.20. Restaurant.

Figure 7.1.21. Cinema.

Figure 7.1.22. Overhead damaged at main
vertical zone.

Figure 7.1.23. Cunard Ambassador exterior
view.

7.1.5 Regulations and Requirements

7.1.5.1 Coast Guard Regulated Vessels

This section addresses the regulatory requirements pertaining to polymeric materials as they apply to Coast Guard-regulated vessels at the national and international level.

In general, the main areas of concern are the combustible nature of structural materials used for bulkheads and decks, and nonstructural materials used for fur-

Figure 7.1.24. Access hole by CG.

Figure 7.1.25. Undamaged lounge.

nishings. Additional areas of concern are the combustible nature of materials used for deck coverings, passageways and stairways, interior finishes, furnishings, reefer spaces insulation, cargo, hold insulation, and tank insulation.

A historical background pertaining to the evolution of the national and international developments regarding "safety of life" that resulted from the 1929, 1948, 1960, and 1974, Safety of Life at Sea Conventions and the 1977 Fishing Vessel Conference is contained in Appendix A.

7.1.5.2 National

The three specific subchapters, within Title 46 CFR, that apply to the use of polymers presently being used on commercial vessels are as follows:

Marine Engineering Regulations

Subchapter F is a summary of the requirements for materials, construction, installation, inspection and maintenance of boilers, contained pressure vessels, appurtenances, piping, and welding and brazing. (Title 46 CFR, Parts 50 to 63 inclusive). For those vessels affected, some examples of polymeric materials which may be utilized are specifically detailed and their use controlled as indicated below:

MATERIAL	USE
Polyvinyl Chloride (PVC) and Fiberglass Reinforced Polymer (FRP)	Piping for non-vital fresh and sea water service

Flammability Requirements

(1) Self-extinguishing by ASTM D635.
(2) Piping is enclosed or boxed in steel.
(3) Penetration of fire resistant bulkhead controlled by design

MATERIAL (Continued)	USE
Urethanes, Neoprenes, Polyamides, and Butyl Rubber	Nonmetallic flexible hose

Flammability Requirements

Self-extinguishing by test method ASTM D 1691; must be capable of withstanding a free-burning hydrocarbon fire for 2½ minutes under its rated pressure.

Of particular interest is the continued use of ASTM D635 and ASTM D1692 as a criterion of lack of flammability in the applications discussed above. The well-established deficiencies of both test methods (Hindersinn, R. R., 1969–1972) would appear to require a research program to reevaluate these test specifications as a measure of polymer fire retardance. Thereafter, the requirements can be updated using current test methods.

Electric Engineering Regulations

Subchapter J is the standard by which electrical installations are designed, installed, and inspected. In general, it also lists the construction requirements of electrical components. Among the polymers available for use as electrical cable insulation, the most commonly used material is polyvinyl chloride. Currently, all shipboard electrical cable must meet the requirements and fire test provisions of IEEE Standard 45 with reference to IEEE Standard 383. The adequacy of this flammability standard is discussed in more detail in Section 7.4.

Miscellaneous Specifications

Subchapter O contains a wide variety of performance specifications for specific items of required equipment and construction materials. It extends from life-boats to industrial materials utilized for interior finishes. The basic specifications which prescribe performance requirements for materials with respect to fire safety are listed in Table 7.1-7.

7.1.5.3 International

Controls were placed on the use of polymer materials in the 1929, 1948, 1960, and 1974 SOLAS Conventions and at the 1977 Fishing Vessel Conference. (See Appendix A for SOLAS Conventions).

7.1.5.4 The 1977 Fishing Vessel Conference

In March 1977 the first major international conference concerning safety of fishing vessels was held in Spain. While fire safety was not the primary area of emphasis, there was a recognized concern for increased polymer usage in the construction of fishing vessels. One of the final recommendations was entitled

Table 7.1-5. Mishap Statistics

Marlin No. 3	Jackup	Marlin Drilling Co.	Partially submerged while moving to location in Gulf of Mexico. Repaired and returned to service.	1.7
Santa Fe Explorer (formerly Orient Explorer)	Jackup	Santa Fe (formerly Royal Dutch/Shell)	Damaged in Mediterranean Sea while under tow from Borneo to England. Repaired and returned to service.	1.5
Triton	Jackup	Royal Dutch/Shell	Damage caused by blowout and fire in Nigeria. Not salvaged.	1.5
Bruyard (Sedco 135 B)	Semi-submersible	Royal Dutch/Shell	Broke up in South China Sea while under tow, 13 casualties. Not salvaged.	7.5
Papuro	Jackup	Saipem S.p.A.	Destroyed by blowout and fire in Adriatic Sea, 3 casualties. Not salvaged.	6.0
Maverick I	Jackup	Zapata Off-Shore	Destroyed by Hurricane Betsy in the Gulf of Mexico. Not salvaged.	5.7
1966				
Sea Gem	Jackup	Compagnie General D'Equipments	Collapsed in North Sea while preparing to move, 13 casualties. Not salvaged.	5.6
Roger Butin (formerly Neptune III)	Jackup	CEP	Capsized after moving on location off Cameroon, Africa. Water and hull damage. Not salvaged.	7.0
Mercury (formerly Nola I)	Converted YF barge	Golden Lane Drilling	Capsized and sank during storm off Tuxpan, Mexico. Not salvaged.	1.5
Rig No. 52*	Jackup	Offshore Co.	Leg damage. Salvaged.	0.2
1968				
Julie Ann	Jackup	Dixilyn Corp.	Sank while under tow during storm in Gulf of Mexico. Not salvaged.	4.0
Dresser II (converted to Dresser VII)	Jackup	Dresser Offshore	Capsized on location. Salvaged and returned to service. Refurbished rig valued at $1.5 million.	2.0
Little Bob	Jackup	Coral Drilling Co. (now Fluor Drilling)	Blowout and fire in Gulf of Mexico. Derrick collapsed and rig badly burned. 7 casualties. Not salvaged.	2.0
Ocean Prince	Semi-submersible	ODECO	Destroyed on location by North Sea storm while operating as submersible. Hull broken up. Not salvaged.	7.0
Ocean Traveler	Semi-submersible	ODECO	Minor structural damage during storm in Norwegian North Sea. Sprung leak in one of its two main supporting pontoons. Repaired.	Insignificant
Ocean Viking	Semi-submersible	ODECO	Minor structural damage during Norwegian North Sea storm. Repaired.	Insignificant
Nola III	Drill barge	Zapata Off-Shore	Fire-damage in engine room, several engines replaced. Incident occurred off Sumatra. Repaired.	Unknown
Chaparral	Jackup	Zapata Off-Shore	Lost three legs during Gulf of Mexico storm while under tow to Italy. Repaired and returned to service.	2.0
Unknown	Inland drilling barge	Service Contracting	Sank while under tow in Gulf of Mexico. Not salvaged.	1.5
1969				
Wodeco II	Drill barge	Fluor Drilling Services	Ice damage to hull, mast blew off during storm in Hudson Straits while rig under tow. Repaired.	0.4
Wodeco III	Drill barge	Fluor Drilling Services	Blowout, Red Sea. No damage to rig, but underwater equipment lost.	0.5
St. Louis	Submersible	ODECO	Water damage in engine room from Hurricane Camille. Repaired.	Insignificant
OV-2	Tender	Offshore Co.	Capsized and partially sank during storm in Lake Maracaibo. Not salvaged. Deliberately sunk by owner.	1.5
Estrellita	Jackup (tender assisted)	Offshore Co.	Capsized while under tow in Gulf of Mexico. Declared total loss by insurance company. Salvaged by owner and returned to service.	2.5 (paid by insurance company) 1.9 (for salvaging & refurbishing)

(Continued)

Table 7.1-5. Mishap Statistics (Continued)

Discoverer III	Ship-shape, self-propelled	Offshore Co.	Blowout damage (no fire). Repaired.	0.6
Big 60 & Tender OV-1	Jackup (tender assisted)	Offshore Co.	Slight fire damage from diesel fuel line. Repaired.	Insignificant
Discoverer II	Ship-shape, self-propelled	Offshore Co.	Blowout off Malaysia. Deck hatches left open—minimal water damage. Repaired.	Insignificant
Sanda 1 (formerly Drillship)	Ship-shape	Reading & Bates	Gash in hull when collided with French freighter in Gulf of Lyons—damage slight. Repaired.	0.15
J. W. Nickle	Jackup (tender assisted)	Reading & Bates	Storm damage in Arabian Gulf. Jackup declared total loss. Tender salvaged.	2.5
E. W. Thornton	Catamaran	Reading & Bates	Blowout off Malaysia. No reported damage.	None
Stormdrill III	Jackup	Storm Drilling Co.	Severe fire damage from blowout off Texas, 1 casualty. Repaired and returned to service.	3.5
Transworld 61	Semi-submersible	Transworld Drilling	High wind and rough water damage to legs while moving onto location off South Africa. Repaired.	0.8
Glomar North Sea	Drillship	Global Marine	Severe storm in North Sea moved rig off drill site and damaged drilling equipment. Repaired.	Unknown
Mercury*	Jackup	Offshore Co.	Heavy weather damage. Salvaged.	0.3
Westdrill I	Jackup	Westburne Int'l.	Damaged in storm while in tow off Ivory Coast. Salvaged.	0.5
Constellation	Jackup	Offshore Co.	Sank during North Sea storm while in tow. Not salvaged.	5.8
North Star	Jackup	Offshore Co.	Sustained leg damage while in tow during North Sea storm. Repaired.	Unknown
John C. Marthens	Jackup	Offshore Co. Constructors	Suffered leg damage during storm in Gulf of Alaska. Repaired.	Less than $100,000
George M. Reading	Tender	Reading & Bates	Grounded during Hurricane Camille. No reported damage.	None
Rimtide	Submersible	Rimrock Tidelands (now ODECO)	Blowout in Gulf of Mexico. Salvaged.	Less than $100,000
Mariner I	Catamaran, semi-submersible	Santa Fe	Structural damage to hull during rough weather off Argentina. Repaired.	0.2
Sedco 135G	Semi-submersible	SEDCO, Inc.	Severe fire damage from blowout in Timor Sea off Australia. Repaired and returned to service.	3.5
Mercury	Jackup	Offshore Co.	Damaged in Lisbon harbor. Salvaged.	0.1
Scorpion	Jackup	Zapata Off-Shore	Sank in storm off Canary Islands while in tow. Not salvaged.	2.3
Unknown	4 tenders	Chevron Oil	Damaged in Hurricane Camille. All repaired and returned to service.	Less than $100,000
Rig 20	Inland barge	Rowan Drilling	Destroyed in Hurricane Camille.	$800,000
Rig 14	Inland barge	Rowan Drilling	Minor damage sustained during Hurricane Camille.	Insignificant
1970				
Rig 15	Inland barge	Field Drilling	Destroyed in Hurricane Celia.	$500,000-$1 million
Wodeco V	Barge-shape	Fluor Drilling Services	Drill collars fell from derrick and pierced main deck and bottom of hull. All electrical gear in DC generator room, including generators and switch controls, had to be overhauled. Engines were overhauled and hull was patched.	0.7
Unknown	Inland barge	Kelly Drilling Co.	Blowout occurred with fire damage. Not salvaged.	0.5 to 1
Kenting I	Jackup	Kenting Ltd.	Storm in mid-Atlantic while in tow—structural damage (1/70). Repaired. Sabotaged off Ivory Coast—hull damage (3/70). Repaired.	Total damage for mishaps 0.5 million
Rig 59	Jackup	Offshore Co.	Leg damage (1/70). Repaired. Out of work approximately 12 days. Toppled over while operating off Nigeria (5/70). Towed out to sea and sunk by owner. Not salvaged.	Damage less than 0.2 million (4.0 million) (total loss)

(Continued)

Table 7.1-5. Mishap Statistics   (Continued)

1971

Big John	Drill barge	Atwood Oceanics	Blowout off Brunei. Severe fire damage to drilling equipment. Water became aerated and vessel samk until main deck was 3-4 ft under water, 9 casualties. Repaired and returned to service.	4.3
Endeavor	Jackup	Zapata Off-Shore	Lost top part of leg while under tow in rough seas off West Africa. Repaired.	1.7
Ocean Driller	Semi-submersible	ODECO	Gas blowout off Louisiana. Rig eased off location and abandoned. No fire or damage. BOP stack shimmed closed but didn't stop gas from escaping and bubbling water 20 ft into air.	None
Wodeco II*	Barge	Fluor	Blowout and fire off Peru, 7 casualties. Not salvaged.	4.5
Panintoil II	Jackup	AMOCO-Iran (IPAC)	Damaged by storm on location in Persian Gulf. Salvaged.	2.8

1972

Alta Mar II	Tender	Perforaciones Alta Mar	Sank during storm in Lake Maracaiba. Salvaged.	Less than 1.0 million
M. G. Hulme	Jackup	Reading & Bates	Blowout (no fire), cratering. Rig capsized in Java Sea. Not salvaged.	7.5
Rig 60	Jackup	Transworld Drilling	Blowout in Gulf of Martoban off Burma. Lost at sea. Not salvaged.	10.0
J. Storm II	Jackup	Marine Drilling Co.	Blowout in Gulf of Mexico. Not salvaged.	8.0
Intrepid	Jackup	Zapata Off-Shore	Leg failure in Eugene Island area of Gulf of Mexico. Salvaged.	3.5
Ocean Tide	Jackup	ODECO	Sustained high wind damage in U.K. sector of North Sea. Salvaged.	Unknown
Mr. Arthur	Submersible	Fluor Drilling Services	Major damage in Gulf of Mexico (South Pass, Block 26). Salvaged.	Unknown

1973

Neptune 6	Tender	Forex-Neptune	Struck platform during storm in Persian Gulf and sank. Total loss.	1.0
Mariner 1*	Semi-submersible	Santa Fe	Blowout off Trinidad, 1 casualty. Repaired and returned to service.	Unknown
Topper III	Jackup	Zapata Off-Shore	Damaged in Gulf of Mexico. Under repairs in Vicksburg, Miss.	Unknown
C. E. Thornton*	Jackup	Reading & Bates	Damaged while under tow from Persian Gulf to Red Sea. Total loss.	5.0
Rowan Anchorage	Jackup	Rowan Drilling Co.	Leg collapsed while jacking up in the Macassar Strait off E. Kolimanian. Salvaged.	3.0

1974

Transocean III	Semi-submersible	Transocean Drilling	Capsized and sank in U.K. sector of North Sea during storm. Not salvaged.	20.0
Transworld 61*	Semi-submersible	Transworld Drilling	Started cracking up in Danish North Sea during storm. Under repairs.	Unknown
Dresser VII*	Jackup	Dresser Offshore	Capsized while under tow in Gulf of Mexico, 1 casualty. Not known whether rig will be salvaged – it is lying on its side in 30 ft of water. Mishap under investigation.	Unknown

(Concluded)

Table 7.1-6. Major Aircraft Carrier Fires (Since 1965)

Year	Ship	Cause	Cost ($M)	Fatalities
1966	Oriskany (CVA 34)	MK-24 Flare	10	43
1967	Forrestal (CVA 59)	Aircraft fuel flight deck	20	134
1969	Enterprise (CVAN 65)	Bomb flare flight deck	5	27
1972	Forrestal (CVA 59)	(Flag office)	10	–
1973	Saratoga (CVA 60)	(Electronic area)	5	–
1973	Kitty Hawk (CVA 63)	Oil strainer leak (#1 MMR)	1	6
1974	Enterprise (CVAN 65)	Vast space	–	–
1975	Kennedy (CVA 67)	Collision/stowage	1	1

Table 7.1-7. Major Ship Fires

Year	Ship	Cause	Cost ($M)	Fatalities
1972	Newport News (CA 148)	Gun turret explosion	1.5	21
1973	Roark (DE 1053)	Lube oil (MMR)	–	–
1971	Knox (DE 1052)	Fuel overflow near hot boiler	0.132	–
1973	Force (MSO 445)	Engine room fire	4 (total loss)	–
1974	Enhance (MSO 447)	Lube oil spray explosion and fire	1.6	–
1975	Belknap (DLG 26)	Collision JP-5, Ordnance cook-off	213	7

"Guidance Concerning the Use of Certain Plastic Materials: "In considering the problem concerning the use of certain plastic materials, particularly in accommodation and service spaces and control stations, the Administration should note that such materials are flammable and may produce excessive amounts of smoke and other toxic products under fire conditions."

The use of polymers aboard vessels which must meet requirements laid down by international convention has gone through a cyclical change. The trend is towards the controlled use of polymers. This trend should have a dramatic positive impact on the overall fire safety of vessels in international trade.

7.2 Ship Structural Design

7.2.1 Coast Guard Regulated Vessels

The Coast Guard is involved in the development of fire safe surface ships at the national level and is working hand in hand with the Intergovernmental Maritime Consultative Organization (IMCO) to achieve the same goal at the international level. The Navy has its own ongoing research and development programs in these areas, and formulates its own regulations to achieve the same goals.

The Coast Guard controls the use of flammable materials in shipbuilding by enforcing the requirements contained in the Code of Federal Regulations Title 46-Shipping. All registered vessels must comply with these regulations as mandatory requirement for certification and documentation under the U.S. flag.

Required shipboard structural design and fire protection are elements of an overall "Life Safety System" incorporated into the current vessel regulations. Detailed requirements specify minimum design features which provide for the vessel's:

1. Subdivision and stability
2. Manning
3. Fire protection (active and passive)
4. Lifesaving appliances and cargo containment

Operational requirements, as well as other detailed structural requirements for certification and inspection, are also contained in the appropriate vessel regulations. Due to the unique features of a vessel's environment, the design approach with respect to function, utility, and life safety has to be coordinated into a workable system. The general approach of many maritime regulatory bodies is the specification of basic principles and materials which allow the designer and naval architect flexibility in achieving their goals.

Structural fire protection is an example of a portion of system design, since construction materials are controlled and design guidance and specific requirements are given with respect to arrangement and means of egress.

Coast Guard-regulated vessels also include hydrofoil and hovercraft type vessels as defined in Section 7.1 and in Chapter 9 of this volume. To date, requirements for structural fire protection have been determined by applying the regulations specified in Title 46-Shipping, Subchapters H-Passenger Vessels (46 CFR 75), and T-Small Passenger Vessels (46 CFR 175).

7.2.1.1 Controlled Versus Uncontrolled Use of Polymeric Materials

As with any given set of criteria, there are cutoffs or limitations. While this is true for commercial shipping (subject to reguation and Coast Guard inspection), those vessels which fall outside the existing regulations have access to the uncontrolled use of potentially dangerous polymeric materials. (See Table 7.1-1 for a comparison of statistics pertaining to casualties involving inspected and uninspected vessels).

7.2.1.2 Criteria of Application

The determination of the applicability of the various requirements is a function of the variables of voyage, gross tonnage, number of passengers, and vessel type.

1. Voyage: where the vessel goes determines the severity of requirements. For a voyage in domestic waters, little or no control exists; while for an international voyage minimum governmental regulations exist in support of treaty commitments.
2. Gross tonnage: the measure of the ship's displacement is often utilized as a starting point for many requirements.
3. Number of Passengers: passengers are not members of the ship's crew. In most cases the transport of passengers leads to more stringent fire specifications and control.
4. Type of vessel: the intended service of the vessel (e.g., cargo, tank, or passenger) results in different criteria of fire safety.

The foregoing factors, coupled with the type of propulsion, serve to set the standards for the control of various items of construction on board commercial vessels. The fitting and installation of systems, components, and materials is contained in Title 46, Code of Federal Regulations "Shipping."

The basic vessel regulations detail fire protection, as well as other design and inspection requirements. The stringency of the structural fire protection requirements is a function of the relative hazard that the vessel encounters in its trade and the number of passengers carried, if any.

In each of the inspected vessel categories, there is a common philosophy required to utilize structural fire protection. This philosophy is known as "building in" inherent fire safety. The basic concept of creating a vessel which has an inherent amount of fire safety had its genesis with the group of technical experts that drafted "Senate Report 184" on the fire on SS Morro Castle. The concept of built-in fire safety (noncombustible construction), as distinguished from active (sprinklers) protection, has withstood the test of time.

The requirements for basic vessel divisional structures, bulkheads and decks, are described in Chapter 5 in conjunction with test description.

The tests for noncombustibility, and the "A" and "B" Class divisions are internationally accepted by the countries signatory to the 1960 Safety of Life at Sea (SOLAS) Convention which is discussed in more detail in Section 7.5 and Appendix A.

7.2.2 Navy Vessels

In general, fire protection design requirements, in conjunction with ship structure, are identified in the detailed shipbuilding/overhaul specifications and applicable hull-type drawings. In addition, many amphibious and auxiliary ships are constructed in accordance with American Bureau of Shipping rules of U.S. Maritime Administration specifications.

Structural elements are designed to provide adequate strength and stability for the ship, relative to design load criteria. Since ship designs must first take into account operational requirements such as range, speed, manning, and ordnance, improvements in fire protection through structural design changes have been limited. The net effect has resulted in a compromise between "individual" fire-tolerant compartments and actual design practice.

The importance of maintaining structural integrity during fire situations has been heightened as a result of recent Navy shipbuilding trends to use lightweight aluminum alloys in high-performance hulls, ship superstructures, and joiner bulkheads. For continuing safety of the ship, it is imperative that the structure be continually protected, maintained, and repaired as necessary.

7.2.2.1 Fire-Zone Boundaries

General Navy Requirements. A fire-zone boundary is a physical boundary designed to retard the passage of flame and smoke. Fire zones apply primarily to transverse bulkheads. Exceptions are longitudinal bulkheads in tank wells, as on landing type ships. Surface ships with an overall length greater than 220 feet shall be divided into main vertical zones by utilizing the main subdivision bulkheads (and portions of decks, where the subdivision is stepped) . Fire-zone bulkheads shall be continued from main subdivision bulkheads through the superstructure, stepped as necessary. The distance between fire-zone bulkheads shall not exceed 131 feet. Ship construction drawings shall show the location of each fire-zone boundary and the tightness of all portions of such boundaries. The tightness of a fire-zone boundary shall never be less than "fumetightness."

7.2.2.2 Material and Construction

The Navy requires that fire-zone boundaries of steel hull ships shall be constructed of suitably stiffened steel and made intact with the principal structure of the ship, such as structural bulkheads, decks, shell, and deckhouse. Ships with aluminum superstructures shall have aluminum fire-zone boundaries in the superstructure. Fittings in fire-zone boundaries, except gaskets for watertight and oiltight closures, shall be of noncombustible material. The stringency of the material specifications allows very little polymeric material to be used in hull, bulkhead, and deck construction except for decorative coatings, paint, or insulation.

7.2.2.3 Compartment Testing

The Navy conducts tests on the various shipboard compartments in order to ensure the integrity of the boundary and its ability to perform its functions of preventing the spread of fire, toxic gases, and water.

Two types of tests are used during the construction of ships to demonstrate adequate tightness of the structure from the standpoint of resistance to the spread of flooding, fire and gases.

7.2.2.3.1 Tightness Tests

Tightness tests serve to assure the designed tightness of the structure under reasonably expected conditions of use and loading. The tests are performed by applying water pressure equivalent to the specified design head of the structure.

7.2.2.3.2 Completion Tests

Completion tests demonstrate adequate tightness of a completed compartment of a ship. These tests are performed by applying air or liquid pressure, by hose testing, or visual inspection. Any lack of tightness is detected by observing the drop in air pressure, the leakage of liquid, or visible openings.

7.3 Damage Control Procedures

Regulations pertaining to fire and smoke detectors, fixed and portable fire extinguishers, and firefighting agents used aboard commercial vessels are specified in Title 46-Shipping, Subchapters D-Tank Vessels (46 CFR 34), H-Passenger Vessels (46 CFR 76), I-Cargo and Miscellaneous Vessels (46 CFR 95), J-Electrical Engineering (46 CFR 113), Q-Specifications (46 CFR 161), T-Small Passenger Vessels (46 CFR 181) and U-Oceanographic Vessels (46 CFR 190 and 193). See Appendix C for additional information pertaining to Coast Guard and Navy fire protection objectives and fire extinguishment information.

Although a discussion of firefighting equipment and techniques is not a specific objective of this committee, firefighting as an integral part of shipboard damage control procedures is an important part of a large vessel's shipboard activities, and as such has a direct bearing on fire safety, as related to polymers.

7.4 Materials

7.4.1 Introduction

This section addresses the uses of polymeric materials, in national and international shipbuilding.

Applications of polymeric materials in the past varied with little or no control over their use: present applications have been controlled (Title 46-Shipping, Code of Federal Regulations) to limit the fire hazard and the resulting effects of smoke and toxicity. As new materials and retrofit concepts are developed, they are normally included in the overhaul and repair of older vessels to bring them up to new ship standards.

The use of polymers in ships for decoration, comfort, noise reduction, and, increasingly, to replace functional metal parts is growing rapidly. Proper selection of materials requires full consideration not only of the functional requirement of the part, but also of the fire performance in every realistic environment in which the part may be found. (See Chapter 4).

Information pertaining to uses of polymeric materials by the shipbuilding industry appears in a study performed for the National Shipbuilding Research Program. (Maritime Administration, 1977). One section, "Current and Historical Applications," is of particular interest to this chapter and is reproduced below.

7.4.1.1 Current and Historical Applications

Prior to 1960, marine usage of plastics, including boat hulls, was relatively limited, considering weight. However, the number of applications in many different ship types was significant. The largest use was in passenger and naval vessels. The British pioneered applications in ocean-going ships.

A compilation of noteworthy marine applications up to 1960 is found in Table 7.4.1. Of these, only two were found unsatisfactory — glazing, which scratched excessively, and wash basins which lost their finishes after the use of abrasives or improper cleansing agents. There are now specially coated acrylic (Lucite, Plexiglass) or polycarbonate (Lexan) glazings which are successfully used, as in railroad cars, because they have much greater scratch resistance than the earlier materials. Also, there are now upgraded plastics that are successfully used for wash basins, as long as proper cleansing instructions are followed.

Vent systems made from plastic duct have been evaluated by the Navy (Anon, 1960). While corrosion, weight, and noise considerations led to the use of fiberglass and PVC, such systems were found to be more expensive. Further, lower stiffness required more support. Vent systems using plastic duct are still more expensive, and are used primarily in corrosive locations.

Table 7.4-1. Plastics Usage in Ships up to 1960

General Use – Area/Application	Plastic Used
Living Quarters	
Bath tub	Fiberglass-Reinforced Polyester (FRP)
Shower/toilet	FRP
Table tops	Melamine
Tops for dining area	Melamine
Tile floors	PVC
Light fixtures	Acrylic
Window (glazing)	Acrylic
Port hole box	PVC
Wash basins	FRP
PVC finished bulkhead	PVC
Wash basins	FRP
Piping	
Waste/sanitation piping	FRP
Waste/sanitation piping	PVC
Washdown piping (above deck)	PVC
Cargo line	Glass/epoxy
Propulsion	
Propeller	Nylon
Propeller	Nylon
Propeller	FRP
Wrapping for prop shaft	Glass/epoxy
Miscellaneous	
Lifeboats	FRP
Wheel house	FRP
Wheel house	FRP
Swimming pool	FRP
Swimming pool	FRP
Permanent awnings	Glass/epoxy
Sonar dome	Steel/plastic skin
Insulated hatches	Polyurethane foam
Food provision room	Polyurethane foam
Lining for storage hold	PVC/plywood

Other marine uses for plastics noted since 1960 are summarized in Table 7.4-2.

The above observations, have been confirmed by numerous naval architects, subcontractors, and shipbuilders interviewed for assessing plastics in specific applications. The information obtained in this manner is summarized in the following sections.

7.4.1.1.1 U.S. Shipyards

The most common applications for polymeric materials found by contacts of the committee with eleven U.S. shipyards are summarized below:

Application	Polymers Used
Joiner Bulkheads	Melamine/Marinite
Flooring	Vinyl/Asbestos
Shower Stalls	Melamine/Marinite
Light Diffusers	Acrylic
Mattresses	Polyurethane
Chairs	Naugahyde

Application	Polymers Used
Dresser Tops	Melamine
Table Tops	Melamine
Desk Tops	Melamine
Curtains	Nylon
Life Boats	Fiberglass
Insulation	Polyurethane Foam
Rugs	Nylon, Acrylic

All of these uses are well established in merchant ships with little variation except in the case of joiner bulkheads and stall showers. (A Coast Guard authority has commented that because there are different regulations for different ship types, prudent selections to match the materials to the applications are required and "more often than not the uses indicated would be permissible." For example, the Passenger Vessels Regulations require that "rugs and carpets in stairways or corridors shall be of wool, or other materials having equivalent fire-resistive qualities." (See 46 CFR 72.05-10(o).)

Table 7.4-2. Marine Plastic Uses After 1960

General Use—Area/Application	Plastic Used
Outfitting	
Wheel house (dredger)	FRP
Port light boxes	FRP
Ship funnel cowl	FRP
Ventilation inlets	Acrylic
Hatch covers	FRP
Insulation	Polyurethane foam
Electric power equipment	Phenolic
Electric power equipment	Asbestos/phenolic
Electric power equipment	Melamine/phenolic
Electrical conduit	PVC
Showers and bath	Acrylic
Chairs (upholstery)	PVC
Chair padding	Polyurethane foam
On deck furniture	FRP
Modular bathrooms	FRP
Rope	Nylon, polyester
Ballast pipe	FRP
Ship bearings	Several
Rudder voids	Polyurethane foam
Covered bulkhead	PVC
Propulsion	
Blade	Glass/
Propeller	Polypropylene
(30" diameter)	Foam
Submersibles	
Deep diving capsules	FRP
Submarine rudder	See (a)
Mast fairings	FRP
Barges	
Lash barge hull	See (b)
Safety	
Helmets	FRP
Welding clothing	Nomex

[a] FRP skin/steel spar/foam.
[b] Double walled honeycomb construction—FRP.

7.4.1.1.2 Foreign Shipyards

(N.B. This project did not require visits to foreign shipbuilders. The material reported is based upon a private communication).

The information in Tables 7.4-3, 7.4-4 and the following are based on inquiries in six shipyards in Japan:

Bulkheads: Decorative surfaces for living quarters may be melamine laminate, polyester laminate, vinyl sheet, or simply paint, depending upon the owner's specification. The substrate for the decorative finish may vary from asbestos board or plywood to particleboard.

Ceilings: Usually the same as used for bulkheads.

Flooring (Covering): Vinyl or Vinyl/Asbestos.

Window Frames; Fiberglass reinforced polyester (FRP) is commonly used for port shrouds.

Baths/Showers: All six Japanese shipyards have installed fiberglass units.

PVC Piping: Two of the six shipyards visited used PVC for fresh and chilled water supply. PVC is being used more for drain, waste, and vent systems.

Hatch Covers: Up to 500 mm in FRP.

7.4.1.2 Commercial Vessels

The Coast Guard is working with industry to improve the fire performance of polymeric materials in shipboard use. Of particular interest is a research program to determine the adequacy of construction in the marine environment.

Polymers are used extensively throughout surface vessels for hulls, mechanical items, paint, thermal and acoustic requirements, and furnishings as indicated in Section 7.4.1.2.1. Many classes of both natural and synthetic polymeric material are used in commercial vessels.

The wide spectrum of polymers in use today can be put into several broad categories: mechanical items (gaskets, packings, seals, flexible connections), electrical insulation, thermal insulation, noise suppression, protective coverings (paints, deck coverings), furnishings (bedding, chairs), protective clothing (gas masks, fire fighting suits), life saving equipment (lifeboats, inflatable lifeboats, life vests), and packaging and packing material.

The basis for the choice of a material for a given application is primarily function, availability, and life-cycle cost. Additional factors such as weight savings, user acceptance, and environmental resistance also influence the choice.

The materials used in Coast Guard-regulated vessels are controlled by a variety of regulations which have been developed to limit the flammability of materials used in ship construction. The regulations are summarized in Table 7.4-5 (U.S. Coast Guard, 1974) together with a list of typical materials that meet the regulations. As can be seen in this table, the flammability specifications vary from the very stringent, (e.g., 46 CFR 164-007, -008, and -009 used in division bulkheads, which effectively limits choice to noncombustible materials) to the other extreme where

Table 7.4-3. Plastics Usage Noted in Foreign Shipyards

General Use–Area/Application		Materials Used
Living Quarters		
Bulkhead		Melamine/Asbestos/Melamine
		Melamine/Plywood/Melamine
		Polyester/Asbestos/Polyester
		Polyester/Plywood/Polyester
		Vinyl/Plywood/Vinyl
		Paint/Plywood/Particle-board
Ceilings		As above (*a*)
Flooring		Vinyl/Asbestos
		Vinyl
Window		FAP
Frames		Aluminum
		Painted
Baths (conventional)		Melamine/Asbestos/Melamine
		FRP
		Tile/Steel
Baths (modular)		FRP
Doors (exterior)		Polyester/Plywood/Polyester
		Paint/Hollow Steel
		FRP
		Hollow Steel
Doors (inside)		Plywood/FRP
		Polyester/Plywood
		Melamine/Plywood/Melamine
Doors (corridor) (*b*)		FRP/Steel/FRP
		Fiberglass/Plywood/Fiberglass
		Steel/Paint
Piping		
Fresh	(cold)	PVC
		Copper
		Steel
	(hot)	Steel
		Copper
Sanitary	(flush)	Steel
		PVC
		Copper
	(drain)	Steel
		PVC
Machinery		
Chocking		Steel
Electrical		
Conduit for wires		Steel
		Plastic (PVC)
Heat-Air Conditioning		
Fixture inlets to rooms		Steel
		FRP
Cargo		
Hatch covers		Steel
		FRP
Safety		
Life Boats		Wood
		FRP

[a] Kawasaki (Kobe) uses painted plywood for ceilings.
[b] Door from bridge to corridor.

applications such as furnishings are unregulated, or tank insulation where the less stringent ASTM D 1692 specifications allow all but the most flammable of polyurethane foams to be used. The relative effectiveness of these regulations in providing the desired degree of fire safety on commercial vessels will be discussed later.

144

Table 7.4-4. Plastics in VLCCs

Purpose	Kind of Material	Foreign Ships No. 1	Other Foreign Ships No. 2
Rudder bearing bushing	Phenol resin	100 kg	100 kg
Fresh water pipe Sea water pipe in Potable water pipe living Scupper pipe quarters	PVC pipe (including fittings)	2.2 tons	2.2 tons
Flooring tile (cabin, passage)	PVC/Asbestos	600 m^2	1,300 m^2
Deck composition (weather deck)	Synthetic rubber (Neoprene)	200 m^2	750 m^2
Deck composition (in accommodation space)	Synthetic rubber	1,870 m^2	1,900 m^2
Overlay of wall surface	Laminated hard plastic, PVC, cloth	4,000 m^2	11,000 m^2
	Melamine	–	1,700 m^2
Door of sanitary space	FRP/Ply/FRP	15	5
Skirting plate of cabin wall	PVC (Recessed)	–	700 m
Handrail, Stormrail	PVC/Steel	–	0.5 t
Reefer door	FRP with urethane for heat insulation	4 sets	4 sets
Reefer insulation	Foam urethane	1 t	1.3 t
Cover for insulation	Canvas with polyester overlay over glass wool	750 m^2	400 m^2
Chair	Vinyl covered, with polyurethane stuffing	135	120
Sofa	Vinyl covered, with polyurethane stuffing	55	45
Mattress	Cotton cloth urethane stuffing	55	45
Desk, table	Laminated hard plastic top (Melamine)	60	55
Lifeboat	FRP	–	2
Life raft	FRP	–	2
Awning (bridge, swimming pool)	Corrugated FRP	230 m^2	–
Valve lining (corrosion resistant)	Polyester	abt 100	–
Paste, filler	Various kinds	abt. 3 t	abt. 3 t
Paint	Vinyl	10 t	–
	Tar-epoxy	75 t	180 t
	Epoxy	1 t	80 t
Electric wire	Synthetic rubber and vinyl	abt. 100,000 m	abt. 80,000 m
Compound (for insulation and stuffing)	Silicone and epoxy	abt. 100 kg	abt. 100 kg
Bathroon - modules		2 units	20-25 units FRP 17-22 shower FRP

In addition to the broad spectrum of materials described in this section, Table 7.4-5 lists polymeric materials commonly utilized on board regulated commercial vessels. Each area details whether there are specific requirements for the control of polymeric materials and what types of polymers would be permitted, if any. In addition, a miscellaneous materials section has been added for polymers which do not fit the given categories.

Table 7.4-5. Polymeric Materials Commonly Utilized on Board Regulated Commercial Vessels

Item	Specification	Materials Usually Meeting the Specifications
I. Division Bulkheads		
"A"-Class	46 CFR 164.007–Prevents passage of smoke and flame for 1 hour. Noncombustible by test (46 CFR 164.009)	Steel suitably stiffened, minimum thickness is 11 gage
"B"-Class	46 CFR 164.008–Prevents passage of flame for 30 minutes, thermal insulating valve required in certain configurations	Asbestos cement board Gypsum board faced with steel Mineral wool, reinforced with steel Mineral wool, reinforced with laminated plastic Aluminum faced mineral wool
"C"-Class	46 CFR 164.009–Noncombustible	All noncombustible metals, composites of aluminum and fiberglass
"A"-Class Divisions with thermal integrity requirements of 15, 30, or 60 minutes	Base component of division materials added to composite structure must be noncombustible (46 CFR 164.009) and be either a 46 CFR 164.007 or 164.008 class material	Steel (min. 11 gage) Mineral wool Asbestos cement board
II. Decks		
Deck without required fire resistance	N/A	Any gage steel
Decks with required fire resistance	46 CFR 164.006 "Deck Covering" Base material steel (11 gage) Also 164.007 "Structural Insulation" and 164.008 "Bulkhead Panels"	Minimum 11-gage steel plus an overlay of materials such as magnesite, portland cement (maximum of 12% organic content, often wood fibers or polystyrene beads)
III. Deck Overlays		
Overlays	If the overlay is 3/8 inch or less no specific requirements	Asphalt tile Linoleum* Urethane Neoprene Latex Vinyl tile Vinyl asbestos
IV. Linings (interior finish)		
Interior finishes	Maximum thickness allowed 0.075 inch (Vessel Regulations) 46 CFR 164.012 "Interior Finishes" E-84 tent: Flame Spread $\leqslant$20 and Smoke Generation $\leqslant$10. Paint may not be nitrocellulose based.	Paint–Oil or latex based Vinyl coated fabrics Laminated plastics (melamines or porcelain)
V. Passageways and Straiways (fire resistant furnishings required)		
Furniture (case)	Non-combustible materials	Steel, aluminum
Furniture (free standing)	Frame (non-combustible material) Upholstery–46 CFR 164.011 Padding–ASTM D1692 or non-combustible material	Steel, aluminum Neoprene
Carpet	Wool or Equivalent, E-84 test: Flame Spread $\leqslant$75 and Smoke Generation $\leqslant$100	Wool, acrylic, nylon, and fiberglass
Draperies	46 CFR 164.009 Non-combustible materials UL 703, NFPA 701 or 164.001	Fiberglass, treated polyester blends
VI. Furnishings (where fire resistant furnishings are required)		
Furniture (case)	Non-combustible material 1/8 inch combustible veneer top permitted	Steel, aluminum
Furniture (free standing)	Frame–non-combustible Upholstery–no requirement Padding–no requirement	Steel, aluminum Fabric, vinyl Urethane, flexible
Bedding	Sheets–no requirement Mattress–no requirement	Cotton Cotton polyester blends
Carpet	Wool or Equivalent E-84 test: Flame Spread $\leqslant$75 and Smoke Generation $\leqslant$100	Wool, acrylic, polyester to levels
Draperies	46 CFR 164.009–non-combustible 46 CFR 164.011–tentative test method for fabrics	Fiberglass, treated cottons

(Continued)

Table 7.4-5. Polymer Materials Commonly Utilized on Board Regulated Commercial Vessels

Item	Specification	Materials Usually Meeting the Specifications
VII. Furnishing (where furnishings are not controlled) (Notes 1 and 2)		•
Furniture (case)	No requirements	
Furniture (free standing)	No requirements	
Bedding	No requirements	
Carpet	No requirements	
Draperies	No requirements	
VIII. Reefer Spaces (Refrigerated Spaces) (Note 3)		
Accommodation and service spaces insulation	ASTM D1692	Urethane, mineral wool, and fiberglass batts
Cargo holds insulation	ASTM 1692 (suggested)	Urethane, polystyrene, mineral wool, and fiberglass batts
IX. Cargo Area (tankships or liquified gas carriers)		
Tank insulation	ASTM D1692	Urethane
	No requirement if cargo hull is inerted	Fiberglass reinforced plastic composites
Pipe insulation	ASTM D1692	Urethane
X. Piping Systems (plastic)		
Nonmetallic pipe	46 CFR 56.10-5	Polyvinyl chloride (PVC)
Nonmetal service (fresh and saltwater)	46 CFR 56.60-25(a) (ASTM test specs.: Pipe ASTM D1785 and D2241; and, ASTM D2464, D2466, and D2467)	
Vital service	46 CFR 56.60-25(b)	
Flexible hose	46 CFR 56.60-25(c)	
Valves, fittings, and flanges	46 CFR 56.60-25(d)	
Short expansion joints	46 CFR 56.60-25(e)	
XI. Miscellaneous Materials (Notes 4 and 5)		
Sea chest screening	MIL-I-21607	Fiberglass reinforced plan
Sun deck awnings and/or supports		
Lifeboat bilge mooring		
Electrical control flooring		
Ship staging		
Pipe guards on deck, in cargo holds, and engine rooms		
Main deck pipeline crossover cutwalls		
Fore and aft main deck cutwalls		
Lifeboats		
Covers for liferafts		
Shipping containers (Note 6)		
Modular showers		

*Not generally used.

NOTES:
1. On passenger vessels, only, the total volume of combustible face, trim, moldings, and decorations including veneer, in any compartment shall not exceed a volume equivalent to 1/10 inch veneer on the combined area of the walls of the compartment. Items of trim such as moldings or decorations shall not perform any structural function.
2. Total fuel loading not to exceed 7.5 lbs/ft^2.
3. Boundaries of the reefer space must be steel and the covering on the interior of the space must be a hard surface capable of being cleaned. Non-combustible materials as well as fiberglass reinforced plastic (no flammability limitations) are often utilized.
4. Fiberglass reinforced plastic gratings may not be used in an escape route from machinery or other spaces. Additionally for those vessels fitted with a deck foam system, access to the foam firefighting stations must be of steel or equivalent material.
5. If modular FRP showers are utilized the total fuel loading of the compartment must not exceed 7.5 lbs/ft^2.
6. Intermodal shipping containers are not required to meet mod. spec. fire reference requirements.

7.4.1.2.1 Hull Construction

Hull Material. In early days, fires were frequent and often disastrous. Some idea of the seriousness of shipboard fires can be realized by examining the records of steamboats operating in the Nineteenth Century. In the period from 1811 to 1851, fires accounted for 17 percent of the total accidents and 21 percent of the estimated property loss among steamboat accidents. For the decade of 1870 to 1879, fires were second only to collisions as the cause of marine accidents, being responsible for 334 incidents. In this same period, fires caused the greatest number of fatalities, 563. Explosions resulted in another 116 accidents, causing 562 deaths (Hunter, L., 1949).

As late as 1930, some U.S. passenger ships were still being built primarily of wood, although steel had supplanted wood as a hull material. The usual construction technique was to construct the exterior of the vessel of steel and to frame the interior of wood much as one constructs a brick veneer building.

With the introduction of steel as a hull material, some efforts were made to reduce the fire danger of predominantly wooden vessels. The 1929 International Convention for the Safety of Life at Sea required passenger ships to be subdivided by steel fire resisting partitions, spaced at a maximum of 40 m (131-foot) intervals, and capable of resisting a 1500°F fire for 1 hour. Although this standard represented major improvements, it was deficient in a number of respects: it did not specify suitable testing procedures; it did not call for suitable closure of openings; and it did not recognize that the total available (polymeric and natural) fuel supply needed definition and quantification. In addition, various provisions of Title 46 "Code of Federal Regulations-Shipping" require that the hull, decks, and deckhouses of merchant vessels be constructed of steel or equivalent material. The equivalence clause permits in some instances the utilization of thermally insulated aluminum. Together these regulations have effectively eliminated the use of combustible materials in the construction of hulls and bulkheads of large commercial ships. Exceptions to Title 46 regulations have been allowed for certain types and sizes of vessels, traditionally constructed of wood. Such vessels are permitted to be constructed of Fiberglass Reinforced Plastic (FRP). FRP has been extensively utilized in small passenger vessels which carry more than 6 but less than 150 passengers.

In 1972 it was recognized that FRP hulls constructed of general purpose resin presented a potentially serious fire hazard. A detailed analysis of casualties as well as an analysis of the basic material was conducted. From the outset it was recognized that wood always would be largely used in the construction of this type of vessel; therefore, the fire hazard properties of wood were utilized as a benchmark for the development of requirements for FRP hulls. Three specific parameters were investigated: ease of ignition, spread of flame, and heat of combustion of the material.

It was found that hulls constructed of fire retardant resin to comply with MIL-I-21607 closely approximated the fire hazard properties of wood. Detailed regulations were drafted which require that this type of vessel, after a certain date, use MIL-I-21607 resin if FRP is utilized as a primary structural material.

Although the fire safety of small passenger vessels was significantly improved by these regulation changes, the use of more fire retardant polyester resins (those having flame spread ratings significantly less than that of wood, i.e., less than 25 rather than the 100–200 currently allowed by MIL-I-21607) would be expected to further improve the fire safety of these FRP vessels. This subject is discussed in greater detail in Chapter 10.

The military specification only applies to polyester resin. FRP can be made with other kinds of resin; hence, as a minimum, these FRP must possess a degree of fire safety equivalent to MIL-I-21607.

7.4.1.2.2 Structural Fire Protection for Large Commercial Vessels

Structural materials used in the construction of bulkheads and decks of large commercial vessels are regulated by provisions of Title 46, Code of Federal Regulations as specified in Subchapters -I (46 CFR 92.05 and 92.07-50) and -H (72.05) and CG-190 Equipment Lists. 46 CFR 72.05 identifies various structural fire protection classifications that are required for steel bulkheads and decks in the various part of the different classifications of vessels. It identifies in detail the type, thickness, and relative positions in the structure of insulation, panels, and coverings necessary to meet Class A-60, A-30, and A–15 requirements. (See Table 7.4-5 for details of bulkhead and deck specifications). Equipment List CG-190 identifies the manufacturers of approved insulation panels and noncombustible materials that can be used to construct bulkheads or decks meeting the desired fire protection classification. Navigation Vessel Inspection Circular NVC 10-63 also contains sketches of the arrangements of bulkhead and deck constructions used on commercial vessels.

The use of plywood for nonstructural interior bulkheads in the superstructure on existing ships which were originally certificated as cargo or miscellaneous vessels is acceptable provided the requirements of 46 CFR 92.05 or 92.07-90 are met. It is realized that there are other materials which would minimize the fire hazard as per 46 CFR 92.05-1(a) and 92.07. These are recommended. However, there are no requirements which prohibit the above mentioned use of plywood on existing cargo vessels. Plywood is prohibited, however, for non-structural interior bulkheads in similar locations on passenger and tank vessels, and on all cargo and miscellaneous vessels of 4,000 gross tons and over contracted for after January 1, 1962.

These complicated and detailed regulations, which will not be discussed in detail here, are the basic fire prtoection guidelines of the U.S. shipbuilding industry. The regulations are so defined that little or no flammable polymeric materials are allowed in the basic structure of the vessels to which they apply.

7.4.1.2.3 Noncombustible Materials Required in Construction
of Living Spaces on Tank Vessels

The Rules and Regulations for Tank Vessels (CG-123) describe in general language the requirements governing the construction of the living spaces on tank vessels (i.e., the staterooms, hospital spaces, passageways, public spaces such as messrooms and recreation rooms, and similar spaces). The requirements in 46 CFR 32.60-25 and 32.40-1(d) (U.S. Coast Guard, 1977, 1978) call for the use of "fire resistive materials" for the constructions and insulation of all living spaces on tank vessels. The term "fire resistive materials" as used in these regulations means a "Noncombustible Material" as approved under the specification in 46 CFR 164.009 and listed in the pamphlet "Equipment Lists," CG-190. It is intended that within the living spaces, all materials of construction, including panels and insulation, together with any materials used in their erection or for their support, shall be approved "Noncombustible Materials." The only combustible materials of construction permitted within the living spaces are decorative veneers and trim on the panels of staterooms and public spaces. No combustible materials are permitted in the passageways or in hidden spaces. There is no restriction on the type of furniture or furnishings to be used.

7.4.1.3 Fire Loading

One of the most important and an often neglected factor is fire loading. The fire load is the expected or maximum amount of combustible material in a given area. In commercial and military vessels, this consists of the combustible structural and non-structural elements and the combustible contents contained within a single fire area (control spaces, escape routes, accommodation spaces, service spaces, and main and auxiliary machinery spaces). Fire load is usually expressed as weight of combustible material (expressed as equivalence to wood) per square foot of fire area. An evaluation of the fire load within a fire area is a good basis on which to calculate the strength of required fire barriers (divisional bulkheads, doors, floors, etc.). This type of calculation offers a realistic basis for evaluating the amount (thickness) of insulation required to protect steel and aluminum structures for a predetermined period of time.

A typical fire loading calculation to determine the amount (thickness) of insulation required to protect the structural integrity of aluminum bulkheads is in Aluminum Fire Protection Guidelines, SNAME T&R Bulletin 2–21, dated July 1974, [Maritime Administration, 1977] (See Appendix D).

This approach for determining fire load was undoubtedly sufficiently accurate when most shipboard flammables were wood or cellulose derivatives, all of which exhibited a heat of combustion in the range of 6,500 to 8,800 BTU/lb (3600 to 4900 cal/gm). Such a procedure could lead to a serious underestimation of the fire load in a compartment when synthetic polymers replace wood and cellulose as the main combustibles because of the much higher heat of combustion of most hydrocarbon polymers (the range is 16,000 to 20,000 BTU/lb); 8,900 to 11,000 cal/gm. This means that the present procedure could lead to an underestimation of the fire load by a factor of 2 to 3, depending upon the ratio of polymeric material to wood included in the total of compartment combustibles. Some correction or modification of this calculation is necessary if serious underestimation of fire hazards is to be avoided. Perhaps the fire hazard is better calculated on BTU/ft^3 of the room.

7.4.1.4 Accommodations and Service Spaces

As indicated in Tables 7.4-1 to 7.4-4, the polymers used most extensively in these spaces are fiber reinforced polyester, melamine acrylic, PVC, urethane foam, and plywood. The flammability for these materials can vary from very low for melamine acrylic to very high for polyurethane foam. This wide variation in flammability is a result of the wide variation in use of flammability tests to control materials used. The following sections discuss these materials problems in more detail.

7.4.1.5 Habitability

Habitability is an all-encompassing term to describe housekeeping amenities which make for a more comfortable living atmosphere aboard ship. Among these items are carpeting, carpet padding, privacy curtains, mattresses, upholstered furniture, vinyl tile deck coverings, wallboard, insulating materials, and bulkhead and overhead sheathing. Many of these materials, under current specifications, ignite easily and produce large volumes of toxic smoke. (See 5.3.2 for reference to U.S. Navy Habitability Guidance List).

7.4.1.6 Interior Finish

Paint has been a traditionally acceptable interior decorative coating. However, in modern vessels, ship owners and seamen want a greater variety of colors, designs, and patterns. The traditional mahogany and teak wood interiors have, through concern for fire safety, been replaced by wood veneers, high-pressure melamine laminates, and vinyl wall coverings. The maximum thickness of any combustible interior finish is limited to .075-inch. This effectively serves to limit the total amount of combustibles in any single space. By international treaty certain interior finishes are required to be of the "low flame spread" type. From test data this means the low flame spread type of finish must have a flame spread rate not greater than 20 and smoke generation not greater than 10 when tested in accordance with ASTM E-84. The areas common to all vessels that must comply with structural fire protection requirements are the corridors, stairways, stairtowers, and hidden spaces.

7.4.1.7 Furniture

Furniture is essentially divided into two primary categories: "fire resistant" and "non-fire resistant." The term "fire resistant" is a misnomer, in that only the frames are required to be metal; the padding and upholstery only have to be "self-extinguishing" when tested in accordance with ASTM-D-1692, whose limitations are discussed in Section 5.2 of Chapter 5. It should be pointed out that fire resistant furnishings are required in all habitability areas aboard all vessels.

7.4.1.8 Deck Covering

In order to attain the horizontal thermal integrity requirements for divisional structures, as required by regulations, numerous types of flexible deck covering materials have been developed. Due to the need for greater flexibility of the material and considering the lesser impact that floor coverings have in a fire situation, a maximum organic content of 12 percent is allowed for this type of material. Magnesite is often used as a material base. The composite material must then meet a thermal endurance fire test similar to that described in ASTM E-119.

7.4.1.9 Carpets

Carpets are most often found on passenger vessels. International convention requires the carpets to be wool or the equivalent in fire resistance. What "equivalent" means is a direct fire consideration. The Coast Guard utilizes a value of 100 for smoke and 75 for flame spread when tested in accordance with ASTM E-84.

7.4.1.10 Engineering Spaces

Polymers are used in engineering spaces largely as soundproofing, piping, and electrical cables. Minor applications for polymers in these spaces are gaskets, pipe covering, packings, seals, flexible connections, etc.

7.4.1.11 Electrical Cables

The largest single use for polymers in the engineering spaces is as electrical insulation on a host of different electrical cables and wiring necessary for the functions of a modern ship. These cables are covered with wire braid for physical protection and are heavily insulated with plasticized polyvinyl chloride (PVC), which, when heated, produces a dense toxic smoke, and which can carry fire from one compartment to another.

Currently all shipboard cable aboard Coast Guard-regulated vessels must meet the requirements of IEEE Standard 45 with reference to the fire test provisions of IEEE Standard 383. (Refer to Title 46-Shipping Subchapter J (46 CFR 111.60) for additional information). This test utilizes a ribbon gas burner at the bottom of a vertical tray. The gas burner gives off 70,000 BTU/hr at approximately 1400°F. Numerous revisions have been proposed which would raise the test thermal flux to 110,000 BTU/hr.

The current regulations appear to be adequate for control of the flammability of large cable insulation and cable fires on commercial shipping. The adequacy of the flammability regulations covering small communications wire is somewhat less certain, and may require revision to reduce the fire hazard. This is especially significant because of the tendency to include these wires in the main cableways in the interest of economy. The fire behavior of such composite cables is uncertain and may be significantly different than that of the individual units as tested.

Fire safety research, as described in Section 7.4.1.17, defining new fire resistant construction at cable penetrations of compartment boundaries, is aimed at increasing the fire safety of cable insulation by preventing fire progression along cables outside the compartment boundaries. This research should be vigorously pursued. The absence of any smoke or toxicity requirements for cable insulation could represent a significant materials problem for the future because of the serious deficiencies of polyvinyl chloride in smoke evolution and the toxicity of its combustion byproducts. The present absence of low-cost fire retardant halogen-free elastomers will make it very difficult to find a suitable replacement for polyvinyl chloride should future smoke and toxicity regulations eliminate the use of PVC as an elastomeric wire and cable coating. A potential solution to this dilemma under investigation by the Navy is discussed in Section 7.4.2.5.

7.4.1.12 Soundproofing

The material commonly used to soundproof the engineering spaces (engine room, machinery, pumproom, etc.) is fiberglass which is attached as batts to the inner surfaces (bulkheads and ceilings) of these areas. There is no specific test requirement for this particular application. However, 46 CFR 164.009 for noncombustible material is unofficially used in the absence of a specification.

7.4.1.13 Piping

The nonmetallic material used for piping is polyvinyl chloride (PVC) and is described in Title 46 Subchapter F-Marine Engineering [46 CFR 56.10-5(d) and 56.60-25(a—f)].

At the present time, PVC pipe may be used for water, nonflammable chemicals, and air service, provided the pressure is limited to 150 pounds per square inch gage and the temperature does not exceed 140°F. Other types of plastic pipe may be used in this service, provided they demonstrate suitability for the intended service conditions and are used within the manufacturer's recommendations for pressure and temperature. Materials such as glass-reinforced resins, or other plastics, may be authorized by the U.S. Coast Guard Commandant if full physical and chemical description is furnished. Flammability of the product shall not exceed the burning rate of the self-extinguishing type, as determined by the standard test method ASTM D635.

Plastic pipe may be used for nonvital fresh and salt water service subject to the limitation imposed by 46 CFR 56.60-25(a). One of the limitations specified is that PVC pipe and fittings shall comply with the following ASTM specifications:

       (a) Pipe (PVC)
             ASTM D1785 (Schedule 40, 80, 120)
             ASTM D2241 (Standard Dimension Ratio).
             Type I Grade 1 or 2.
             Type II Grade 1.
             Type III Grade 1.

>(b) Fittings (PVC)
>>ASTM D2465 (Schedule 80 threaded).
>>ASTM D2466 (Schedule 40 socket).
>>ASTM D2467 (Schedule 80 socket).
>>Type I Grade 1 or 2.
>>Type II Grade 1.

Additional areas of use are as follows: vital fresh and salt water service (56.60-25(b)); nonmetallic flexible hose (56.60-25(c)); plastic valves, fittings and flanges (56.60-25(d)); and short nonmetallic expansion joints (56.60-25(e)) as tested by ASTM D1692.

7.4.1.14 Cargo Spaces

Although a wide variety of polymeric materials are used in various applications in shipboard cargo spaces, they are used most extensively as deck coverings or coatings and as insulation for pipes, tanks, and refrigerated areas. (See Table 7.4-5). The relatively rapid replacement of nonflammable insulation such as fiberglass batts and mineral wool by polyurethane foam has been facilitated by the superior properties, economics, and insulation values of these highly successful polymeric foams, together with the noncritical flammability requirements set for these materials (ASTM D-1692) (see Table 7.4-5).

As currently used, as insulation in cargo spaces, polyurethane foam is sprayed onto cargo space surfaces and allowed to foam in place. As discussed in more detail in Section 7.4.1.15 in Table 7.4.5, and in Section 4.6 of this volume, such foam can pose very serious fire, smoke, and toxicity hazards when the surface is unprotected by some non-flammable material such as steel or aluminum. The absence of regulations requiring such protection in present Coast Guard specifications is one of the most serious breaches of fire safety uncovered by this present study. This problem is especially serious in tank and LNG ships and is discussed in more detail in subsequent sections on these vessels and in Section 7.4.1.2.6.

7.4.1.15 Thermal Insulation

The polymeric materials most commonly used for thermal insulation aboard ships (cargo, tank, passenger, special purpose, etc.) are the rigid and flexible polyurethane foams. These materials have excellent insulation and sound absorption qualities but low thermal stability which increases their flammability potential. The most common use of rigid foam is as insulation for tanks and cargo holds, piping, reefer spaces (enclosed between metal sheets), and also as a sound barrier. This material is used quite extensively for insulation aboard LNG tankships and in limited quantities aboard other ships. Polyurethane foams are considered to be relatively safe when used covered and a high risk when used uncovered. (When used as a sound barrier they are left uncovered). The flexible foams are used in a variety of applications, the most common being in refrigerated holds or galley reefer spaces. Refer to Section 4.6 and to Table 7.4-5 for additional information pertaining to uses and flammability and toxic smoke production of polyurethane foams.

As indicated in Table 7.4-5, low density polymeric foams are and may be used in large quantities in a variety of shipboard applications, e.g., as padding in fire resistant furnishing, mattresses, insulation and sound deadening. All of these applications are allowed by current regulations even in fire resistant furnishings with the only flammability requirement, if any, being conformance with ASTM D-1692. Although Note 2 of Section VII of Table 7.4-5 limits the total fuel loading of any furnished area to 7.5 lbs of flammable material to the square foot of floor space, this regulation assumes a heat of combustion equivalent to wood or cellulose and makes no distinction among flammable materials in this respect. As indicated in Table 1 of Chapter 9, however, the heat of combustion of such polymers as polystyrene, urethane, or ABS is 2 to 2.5 times that of wood resulting in a fuel loading 2—2.5 times that of the equivalent amount of cellulose upon which these regulations were based. It would appear that prudence would require that this regulation should be carefully reassessed in the light of the increasing substitution of polymer for wood.

Another regulation of doubtful validity as regards fire safety is Note 3 under Section VIII, Table 7.4-5. This regulation requires that spray-applied urethane foam must be bounded by steel at the periphery of each compartment, but requires only that the surface of the foam facing the main cargo storage area have a "hard surface" capable of being cleaned. This regulation thus allows unprotected spray-applied foam to be the main wall surface of such reefer spaces, so long as the natural foam skin is capable of being cleaned. It is just such an unprotected foam surface that has been found hazardous under the Federal Trade Commission consent decree regarding urethane insulating foams. It would appear that such a fire regulation should be carefully reassessed in view of the demonstrated fire hazard of such construction.

The inadequacy of the ASTM D-1692 test as a measure of flammability in any real sense has been demonstrated (see Volume 2 of this report) by several research groups and its use for this purpose is being discontinued. Foams designated as self-extinguishing by this test, when used in large-scale commercial applications, have led to several disastrous fires, the most graphic illustration being the fire that occurred in an empty petroleum storage tank while the foam was being spray-applied. This resulted in the loss of several lives when the fire spread across the unprotected foam surface with remarkable speed. Although some less flammable foams are commercially available, as measured by ASTM E-84, it has been demonstrated that the fire safety of these materials can be brought into acceptable limits only by requiring that the foam be surfaced with a nonflammable skin such as steel, aluminum, or their equivalent. Indeed, it would appear to be necessary for fire safety to require that all such foam insulation be covered in this manner.

Additionally the use of flexible urethane foam as cushioning and padding in both "uncontrolled" and "fire resistant" furnishings should be carefully reassessed, despite the nominal requirement that the total fuel loading not exceed 7.5 lbs/ft^2 of compartment floor. The high heat of combustion (more than twice that of wood) and rapid thermal decomposition of these materials under real fire conditions already have been discussed, and it leads to a fire hazard quite considerably higher than was contemplated with a fuel loading of 7.5 lbs/ft^2 of wood or its equivalent. This regulation should be carefully reassessed to ensure that the potential fire hazard, in view of the demonstrated high fuel value of the polymers being used in these applications, is not allowed to continue.

7.4.1.16 LNG Tankship Application

Currently LNG tankers are being insulated with such materials as cellular glass (multi-layers), cement plaster combinations filled with asbestos fibers and reinforced with expanded metal mesh, polyurethane foam (sprayed or precast), styrofoam (extruded polystyrene), PVC foam, and mineral wool. These materials are used to avoid embrittlement of the ship's non-cryogenic structural steel by preventing contact with liquefied gas at its boiling temperatures as low as $-162°$C. Together with the need that the inner tank liner stay liquid-tight despite unusual mechanical loads such as cryogenic thermal shock, static loads from the cargo pressing against the inner liner, and dynamic loads during transport, this requirement has made polymeric foams the preferred insulation material because of their high insulation value and their resistance to shock and dynamic stress. The high flammability of the large amounts of foam used has created a serious potential fire hazard, but, there are no suitable alternatives available at this time. Programs are underway to determine methods of minimizing the hazard. It would be necessary that the fire safety of such construction be carefully reassessed.

7.4.1.17 Research on Fire Performances

A substantial program of fire safety research is supported by the Coast Guard, in-house, at the National Bureau of Standards, and at academic institutions. Although a large portion of this program includes research in fire prevention and suppression not germane to this study, determining the fire safety of polymeric materials is a significant part of the program. Examples of this type of fire research are the study of the flammability of glass reinforced polyester (GRP) piping and the fire resistance of cableway compartment penetrations (Sheehan, 1972).

7.4.1.18 Flammability of GRP (Glass Fiber Reinforced Polyester) Hulls

In the spring of 1971, the Safety Equipment Branch of the Office of Merchant Marine Safety of the U.S. Coast Guard initiated a small research program to investigate specific aspects of the fire problem involving GRP in boat hulls. Three parameters were examined: ease of ignition, flame spread, and total heat of combustion. These are considered to represent an adequate measure of fire hazard of combustible materials. Smoke development and its related hazards were not specifically examined. The general comment can be made, however, that greater quantities of smoke are produced when GRP hull systems are subjected to fire than are similar hulls constructed of wooden materials.

Results obtained during the program are summarized in Tables 7.4-6 and 7.4-7. The ease-of-ignition test used in this study was developed by the National Bureau of Standards and consisted of the exposure of two test specimens to a gas pilot flame for a period necessary to produce sustained burning. The shortest exposure time resulting in sustained ignition is designated as the ignition time. The data demonstrates the hazard of GRP (prepared from general purpose resins), compared to wood.

Table 7.4-6. Ease of Ignition Test Results (NBS Test Method)

Sample	Time to Ignition (sec)
GRP laminate with fire-retardant resin	120
GRP laminate with fire-retardant resin	100
GRP laminate with fire-retardant resin	87
Red oak	70
Plywood "Marine Exterior" unpainted	96
Plywood "Marine Exterior" painted	92
GRP laminate with general-purpose resin	52
GRP laminate with general-purpose resin	51
GRP laminate with general-purpose resin	50
Pine "high density"	46
Plywood fir	43
Redwood	37
Pine "low density"	34

Table 7.4-7. Results of ASTM E-84 "Method of Test of Surface Burning Characteristics of Building Materials"

Sample	Average Flame Spread[a]
GRP with fire-retardant laminate	77.2[b]
GRP with fire-retardant laminate	77.1[b]
GRP with fire-retardant laminate	90.3[b]
GRP with fire-retardant laminate	107.0[b]
GRP with fire-retardant laminate	99.0[b]
GRP with fire-retardant laminate	80.2[b]
Good grade of plywood	125.0
Dried red oak	100.0
Assorted hardwoods	60-105.0
GRP with general-purpose resin	400.0

[a] In this test, dried red oak serves as the base level of 100, zero represents no flame spread, and any value greater than zero indicates a higher flame-spread rating.

[b] The values are much higher than need be. The tests should be repeated with Class I materials.

The initial conclusions that can be drawn on the basis of tests conducted to date are that GRP hulls with general purpose resin present more of a potential fire hazard than wood, and GRP hulls with Class II fire-retardant resin represents approximately the same fire hazard as wood.

Although these data were used as a basis for the promulgation of regulations requiring a minimum flame spread rating of 100 for GRP used in boat hull applications, the greater fire resistance of the more highly fire retardant Class I polyester resins, currently available commercially, should have been included in this study for completeness. Such resins exhibit flame spread ratings of less than 25 as measured by ASTM E 84 and could be expected to provide a significant improvement in fire safety in these applications.

7.4.1.19 GRP Piping

Studies have also been conducted in this program to evaluate the flammability characteristics of general purpose GRP as replacement piping in engineering spaces for such materials as steel, copper, and aluminum because of the lighter weight and superior corrosion resistance of the GRP.

To ensure that the safety factor was not sacrificed, sections of GRP piping, copper-nickel piping, and aluminum piping were exposed to liquid fuel fires in a special fire-test chamber. The pipes were first tested dry, then with stagnant water, and finally with flowing water. Under dry conditions, *both GRP and aluminum piping failed in 2 minutes.* With the addition of lightweight protective insulation, the dry pipes remained functional for 8 minutes. The life of the piping was increased only slightly when the pipe was filled with water, but the pipe resisted standard test fire conditions resisted for an hour when the pipe was filled with flowing water.

7.4.1.20 Penetrations

Because of previous experience indicating the ease of fire progression along cableways between compartment penetrations, some general experiments aimed at decreasing fire penetration have been attempted. One approach involved use of a fire-retardant clay-like compound to caulk the holes in compartment bulkheads through which cables passed. This caulking prevented the spread of fire.

In a subsequent experiment, several fire-retardant mastic coatings were prepared for test in the fire chamber. Cables were also painted with the various mastics, then run up the bulkhead and across the overhead of the chamber to simulate shipboard conditions. The setup was then subjected to heat from a hexane gas fire, a relatively clean burning fuel, for approixmately 10 minutes at temperatures up to 815°C. Following a cooling period, the cables were examined and a determination was made as to which material had best withstood the simulated fire conditions. These materials were then applied to protect long cable runs.

In addition to the mastic type coatings, the possibility of wrapping the cables with an aluminum/silicon oxide insulation was studied. The wrapping was then covered by steel mesh. These composite cable coverings were found to be unchanged when exposed to the test conditions described above. They are considered to be the most effectively fire protected cables yet devised. Studies to determine proper installation procedures for these advanced cable coatings are underway.

This data demonstrates the efficiency of fire-resistant compartment penetration barriers in arresting the spread of fire along cableways. It further demonstrates the importance of research in improving shipboard fire safety.

7.4.2 Naval Vessels

The increased use of combustible plastics and other organic materials on naval vessels contributes significantly to the fuel load, and thus to the fire hazard on naval vessels. Materials which carry a label of "self extinguishing" or "flame retardant" often contribute significant quantities of toxic gases and dense, vision-obscuring smoke when exposed to an actual fire environment. A relatively small fire could compromise a mission or result in the loss of the vessel if significant amounts of toxic combustion products (which can cause considerable loss of mental acuity long before they become lethal) are formed. Compounding the problem is the sensitivity of contemporary weapons and weapon systems to damage by heat and corrosive smoke.

The complexity and sophistication of advanced weapon systems like a ship require that almost all components be functioning. A fire that causes abandonment of vital spaces or one that destroys vital electronics and cables severely compromises ship capability. The use of aluminum or composite materials in newer ship superstructures and in the new high performance ships has also increased the hazards due to fire.

The greatly increased costs of ship construction, conversion, maintenance, and repair and the trend towards fewer vessels enhances the value of each ship to the Navy's mission. It also underscores the need for materials with improved fire resistance.

Navy vessels carry large and confined quantities of high-density, high-energy "fuels" and ordnance which are constantly under the threat of unusual ignition sources. High-performance ships, which are becoming more important to the Navy mission, also have special requirements for fire protection because of their construction, reduced manning, and constrained crew mobility.

7.4.2.1 Hull Construction

Standard state-of-the-art hulls are of steel, constructed in accordance with design load criteria identified in the detailed shipbuilding specifications. Special lightweight hulls for high-performance ships (Surface Effect Ships, Patrol Boats, Hydrofoil Missile-Carriers) may require experimental materials such as new aluminum alloys to meet structural requirements. Little if any polymers or oganic composites are now used in naval hull construction; there are no flammability requirements for hull materials. Requirements in force apply primarily to transverse bulkheads (see Section 7.2.2.1) and are not considered pertinent to primary hull structures.

The use of non-metallic materials in other hull construction applications has been limited to such examples as wood for non-magnetic minesweepers, glass reinforced plastics for lightweight ladders, gratings, stanchions, hatches, and fairings, and advanced composites for struts and foils on high-performance ships.

7.4.2.2 Accommodation and Service Spaces

General requirements for use of polymeric materials in the accommodation and service spaces, for ships of the U.S. Navy, are contained in these sections of the General Specifications: 634 (deck coverings), 637 (bulkhead and overhead sheathing), and 640 (living, messing, and recreation spaces). Specific materials used in these spaces are identified in Appendix B.

In general, MIL-STD-1623 is used to set flammability requirements for materials, including polymers, used in these spaces. While this standard generally sets very stringent flammability requirements based upon standard tests such as ASTM E-84 and ASTM E-162, the test for flammability of furniture and deck coverings is based upon FED-STD-501 method 6411 using a unique Navy-approved test procedure. Thermal insulation must meet specificiations determined on a Coast Guard-developed test furnace operated at 750°C.

FED-STD-191, which is the main flammability test controlling furnishings and upholstery, uses a relatively mild vertical bunsen burner test with a 12-second ignition time.

Although both test methods appear adequate for the purpose of determining flammability, the rapid fire progression through the accommodations spaces observed during the Forrestal fire (discussed in Section 7.7.2.1.1) would indicate that there was room for improvement in the control of furnishings and wall coverings. The more stringent flammability standards promulgated as a result that fire have yet to be severely tested by circumstances.

A serious deficiency of the MIL-STD-1623 is the absence of a smoke requirement in FED-STD-501 and FED-STD-191. This omission has already demonstrated its importance in the Forrestal fire, where firefighting efforts were seriously impeded by the large quantities of heavy black smoke from the burning furnishings.

Another serious potential fire hazard allowed by MIL-STD-1623 is in the high ASTM E-84 flame spread value of 250 permitted for light diffusing panels/windows. State-of-the-art materials are currently available for these applications with ASTM E-84 flame spread ratings of less than 25. Such an improvement in flammability specifications would be especially desirable in this case because a relatively high volume of such materials are used in room ceilings and walls, where they are easily melted, drip into the fire below, and contribute additional fuel. The use of less flammable materials should be evaluated relative to the cost/performance improvement expected.

7.4.2.3 Engineering and Cargo Spaces

This section contains Navy shipboard applications and material requirements for thermal, acoustical, and electrical cable insulation used in engineering and cargo spaces.

The requirements for insulation of compartment boundaries, trunks, adducts, machinery, piping, and related equipment are specified in Sections 508, 509, 635, and 638 of the General Specification and MIL-STD-769. (See these references for more detailed information).

Most materials used for the variety of insulation applications, including thermal, acoustic, anti-sweat, and vapor barrier, are largely inorganic in nature (i.e., fibrous glass, calcium silicate, ceramic fiber) and, as installed, are considered to have a low fire risk. However, closed-cell foam (polyvinyl chloride/acrylo nitrile blend), used primarily on chill water piping and selected vertical bulkheads, is required to have a flame spread rating of 25 or less as measured by ASTM E-84. Like some of the serious fire hazards noted in urethane foams that passed similar flammability requirements, the use of significant amounts of this foam without surface protection could cause a serious fire hazard. In addition, the pyrolytic gases from such a foam could be unusually corrosive and toxic. The specification allowing the use of such material in these spaces should be carefully reevaluated.

It should be noted that the use of polyurethane foam for below-deck application is limited to the construction of reefer spaces. In this case, the foamed core is sheathed with stainless steel skins on both sides to provide structural rigidity and thermal protection. This is a well justified precaution in the light of the known flammability of these materials.

7.4.2.4 Thermal Insulation

Thermal insulation is used extensively throughout Navy vessels as insulation on piping, machinery, and compartment bulkheads. In addition, it is used to reduce noise. The thickness of the insulation required for the individual applications varies widely and is controlled by a complicated set of detailed specifications.

According to MIL-STD-1623, such insulation must pass the USCG 164.009 hot tube test, requiring that the sample does not flame, glow, or increase the temperature of the surrounding atmosphere by more than $20°C$, and it must retain at least 50 percent of its original weight. Alternatively, the insulation is acceptable if it passes ASTM E-84 test with a flame spread and smoke rating of zero. In the absence of fire test or hazard data to the contrary, it would appear that these flammability standards are sufficient to preclude the use of the more hazardous polymeric foams in these applications.

7.4.2.5 Electrical Cable Insulation

In general, the applications and requirements for Navy electrical cables are specified in Section 304 of the General Specifications.

For most cable applications, plasticized polyvinyl chloride, neoprene, and silicone jacketing materials provide the required insulating features. More detailed information regarding material construction is identified in MIL-C-915. Cable insulation must conform to the flammability requirements of IEEE 383.

The problem of shipboard fires propagating along cable runs has been addressed by providing fire stops at bulkhead and deck penetrations. Nevertheless, the basic smoke and toxicity questions remain unsolved.

In view of the continuing threat to personnel and vital equipment posed by long and sometimes inaccessible runs, additional efforts are needed to delay and/or limit the involvement of cable jacketing materials in fire situations. Polyphosphazene elastomers present less of a hazard and may provide a solution to the problem. Volume 1 should be consulted for additional details beyond what can be found in Section 4.3.26 and 7.4.2.6 of this volume.

7.4.2.6 Research on Fire Performance Hazard

An overview of Navy-sponsored programs to reduce shipboard fire risks through materials research is presented in the following discussion.

At the Naval Research Laboratory (NRL), the physical-chemical processes of combustion and flammability index-temperature relationships of various liquid hydrocarbons are being investigated to obtain a better understanding of the flammability hazards encountered in the storage, transportation, and handling of hydrocarbon fuels. From these studies have come better insights into the relationships of ignition hazards and sources of ignition and into the methods for the prevention and mitigation of explosions. Also, NRL is investigating toxic agents and aerosols produced by fires in confined spaces to determine the effects of variables such as types of combustibles and reduced oxygen concentration on formation of gases and aerosols, the ability of resulting aerosols to absorb and transport irritant and toxic products, and the role of thermally degradable but non-flammable materials in toxic product formation.

David Taylor Naval Ship Research and Development Center/Annapolis (NSRDC) is studying the degradation characteristics and outgassing products of non-metallic materials in order to assess potential fire hazards and atmospheric contamination. Through these studies, the selection of materials with the lowest level of toxic hazard would be possible. The present work is fundamental in nature and includes both presently used and new candidate materials such as interior paints, adhesives, foams, and electrical insulation.

The National Bureau of Standards, under Navy contract, is attempting to determine the relationship between laboratory controlled fire tests and actual shipyard fires. NBS is examining radiant panel, ease of ignition, NBS smoke chamber, rate of heat release, and potential heat tests on selected materials, then burning combinations of these materials in quarter-scale simulated shipboard compartment configurations. Initial results, as verified by large-scale fire tests, have demonstrated the usefulness of the modeling concept to provide fuel loading limits and to identify potential fire risks.

The Naval Toxicological Unit has initiated studies to determine the toxic effects of compounds on the behavioral patterns (central nervous system disturbances) in primates and lower experimental animals.

Horizons, Inc., Cleveland, Ohio, has developed, under Navy contract, a phosphazene polymer system in which phosphorus and nitrogen form the backbone of the polymer. As a result of this work, it is anticipated that a series of polymers — ranging from liquids and elastomers through rigid plastics — can be obtained. The polyphosphazenes are inherently non-burning and produce less smoke on pyrolysis than more conventional organic materials. These polymers are also intriguing in that they are not corrosive and their eventual costs should be competitive. Current areas of interest include comfort cushioning, electrical cable insulation, and fire retardant coatings. These materials are of special interest, since they are potential solutions to the need for fire retardant electrical cable insulation and fire retardant cushioning for habitability items.

The utilization of habitability materials which are endorsed by the Chief of Naval Operations has resulted in an increased awareness of potential fire risks aboard ship. The Navy's present thrust is to use materials which will fulfill the desired functional service requirements and to avoid the use of those materials which have objectionable fire characteristics as determined by the fire test methods contained in MIL-STD-1623. Finally, the importance of developing and coordinating a comprehensive materials program plan for attacking the fire problem has been recognized, particularly in those areas where the Navy's requirements are unique.

The overall objective of the program is to reduce the vulnerability of naval vessels to the hazards of fire from accidents or hostile action. Ultimately, the goal is to achieve a fire tolerant shipboard environment through materials development, selection, and fabrication techniques.

The principal requirements of the Navy program are to:

- Provide "combustion-proof" materials for critical applications requiring thermal barrier protection.
- Provide "limited flammability" materials that will exhibit improved fire performance characteristics, including reduced flame spread and less production of toxic products of combustion and smoke upon exposure to fire.
- Provide improved extinguishing materials to suppress fires and minimize fire buildup.
- Provide personnel protection materials suitable for firefighting, life support, and rescue operations.
- Develop test methods in support of fire hazard analysis, to enable a realistic prediction of materials behavior in a real fire situation.
- Provide a means of evaluating pyrolysis and outgassing effects of shipboard materials.

7.5 Testing

Testing of polymeric materials is a complex system of test methods, standards, specifications, codes, and related regulations which govern the role and use of polymeric materials. (Please refer to Chapter 5, Sections 5.5 and 5.6, of this volume for a detailed discussion of these tests including recommendations for new test requirements. Further discussion will be found in Volume 2 of this study).

7.6 Smoke and Toxicity

Smoke and toxicity are discussed in Chapter 6 of this volume and in Volume 3 of this study. There are currently no general smoke or toxicity requirements for polymer use on ships. Smoke ratings, as measured by ASTM E-84, are occasionally set as part of MIL-STD-1623B.

7.7 Fire Scenarios as an Aid to Materials Selection and Test Development

7.7.1 Fire Scenarios as a Design Tool in Commercial Vessels

As discussed in some detail in Chapter 3 (and in greater detail in Volume 4 of this series), fire scenarios are beneficial not only as an aid in analyzing specific accidents, but also as a guide to naval architects and marine engineers in material selection and design modifications. In addition, they are helpful in developing realistic fire test methods and standards. Since the current tests and standards for ships were not developed under realistic fire conditions, fire scenarios can guide the formulation of improved regulations and better design.

7.7.1.1 Fire Test on Shipboard Container Construction

This fire-test scenario was developed and analyzed to obtain credible data for a container fire, such as the one that occurred when the SS C.V. Sea Witch and the SS Esso Brussels collided on June 2, 1973 (See 3.4.2). (The lack of credible data for this type of occurrence was a drawback that prompted the U.S. Coast Guard to sponsor and work hand-in-hand with an ad hoc advisory group comprised of fire protection and marine experts from government and industry).

(Refer to Appendix E at the end of this chapter for details of the experiments, results, conclusions and recommendations. Since this report is the work of others, the conclusions and recommendations have not been included in the body of this text).

The findings from this fire test were used to derive the important aspects of container shipping indicated below. From the tests described, it was concluded that an incident resulting in the ignition of cargo within a properly sealed non-damaged container, under most circumstances, would not endanger adjacent containers. However, a container stack exposed to an exterior fire source for more than approximately 5 minutes will most likely ignite and fail structurally, causing the fire to spread to adjacent container stacks. Since external fire exposures produced nearly identical results in all types of containers, it is felt that changes in container construction are not necessary. Containers do not act either to intensify a deck cargo fire or to impede the spread of flame. A fire in an on-deck container stow will spread unless controlled by the installed fire protection system. It is essential that fixed firefighting systems be capable of rapid activation. In Section 7.4.1.2.2, it was pointed out that the cushioning material in the "fire-resistant" category need only pass ASTM D-1692. In 1977, the ASTM Committee under whose jurisdiction this test falls, voted to   remove this test procedure because of its inadequacies, and to substitute a test procedure that would take into account the more rapid flame spread potential of vertical materials.

Similarly, the use of ASTM D-1692 to specify polyurethane foam as insulation must be considered inadequate.

ASTM D-635 is a procedure similar to ASTM D-1692. Its use in the Coast Guard specifications must also be deemed inadequate to determine the fire hazard potential in materials.

Other inconsistencies in test procedures are discussed in Chapter 5, Test Methods, Specifications and Standards, Section 5.2.1, as well as in Volume 2.

The absence of regulations limiting the use of urethane foam, unprotected on the surface by non-flammable skins, is an oversight in Coast Guard regulations that should be corrected at the earliest possible date. Although this omission is especially serious in the absence of adequate flammability standards on the foam itself, the promulgation of more stringent flammability standards should not be considered to obviate the need for non-flammable surface protection. This is true since even foams with flame spread ratings of less than 25 as measured by ASTM E-84 can burn dangerously fast in the presence of a hot ignition source, especially when the foam is in the vertical position, e.g., when in place on the bulkhead in a cargo or engineering space.

7.7.2 Navy Ships

7.7.2.1 Fire Analysis

7.7.2.1.1 USS Forrestal Fire Scenario

During the last 10 years, the Navy has experienced an increased loss in mission capability resulting from accidental fires and arson. Fires on carrier flight decks, in ship machinery spaces, and in combat information centers have provided irrefutable evidence of the serious impact of fire on a ship's ability to remain ready for action.

A notable example occurred aboard the USS Forrestal, CVA 59, on July 10, 1972. In an apparent attempt to delay ship deployment, three individual fires were deliberately set and resulted in damages exceeding $10 million and requiring several months of inactive ship combat status.

The first two fires were discovered at 0145 and 0155, hours, respectively, and involved Class A combustibles. They were both minor in nature and easily extinguished within minutes.

The third fire was detected 10 to 15 minutes later near the War Communication Annex. Vision-obscuring smoke and extreme heat hampered damage control teams in their attempt to isolate the fire. The difficulty in determining exact fire location because of thick smoke complicated securing the appropriate ventilation ducts and, 1½ hours after initial detection, the fire was considered out of control.

No real progress was made in reaching the fire area until about 9 hours later when a hole was cut in the flight deck for hose line access. One hour later, the fire was under control.

Although the source of primary ignition was probably of low order (a cigarette lighter), there was sufficient flammable material present in the Flag Officer's spaces to generate copious quantities of smoke and intense heat, as evidenced by a melted bulkhead and a buckled flight deck. These materials consisted of curtains, carpeting, upholstered and overstuffed furniture, wood paneling on wood furring, paper products, and electrical wire insulation. The actual polymeric materials involved in this stage of the fire are unknown.

Fortunately, there were no deaths and only minor injuries to the base firemen were experienced.

As a result of this conflagration, immediate steps were taken to re-establish control of combustible materials. The identification and removal of all unauthorized, unnecessary, and high-fire-risk outfitting and furnishing materials proceeded on a timely schedule; the development and promulgation of MIL-STD-1623 led to installations of lower fire risk furnishings without sacrificing other desirable habitability features.

7.7.2.1.2 Fire Hazard Analysis — Adequacy of the Method

Although the Forrestal fire, and others, resulted in significant change in fire safety specifications and standards, the scenarios do not appear to have been analyzed fully. Nor have the fire standards and specifications been modified to the extent necessary to prevent a recurrence of the incident.

7.7.2.1.3 Evaluation of Materials and Tests

Navy specifications for test procedures are generally more rigorous than those required by the Coast Guard for commercial vessels; yet specific deficiencies remain, some of which are defined below. A detailed description of those test procedures and their adequacy may be found in Section 5.2.2.

Despite the heavy involvement of the furnishings in the Forrestal fire, FED-STD-191 Method 5903 controlling the allowable flammability characteristics of these materials remains only marginally adequate. Not only does this standard omit any requirement for smoke emission, but also no consideration is made for flaming drip during the flammability test. Both of these omissions are serious and must be corrected before the fire safety control of furnishings can be considered adequate.

Notwithstanding the apparent adequacy of the primary flammability tests on materials other than furnishings used in MIL-STD-1623, the adequacy of some of the smoke standards is doubtful, while the toxicity hazard, as in almost all other fire standards, has not yet been addressed. An example of the less-than-adequate smoke standards is the 450 rating (ASTM E-84) allowed on light diffusing panels and windows. This is a high smoke rating by any current standards. The use of ASTM E-84 as a measure of flammability and smoke without consideration of the dripping characteristic of the polymer under test also requires reevaluation. This is because many thermoplastic foams achieve very low ratings in this test only because they rapidly melt and fall from the source of ignition which is at the top of the tunnel. Since these materials often are not ignited under these test conditions, neither their flammability nor smoke forming tendency is adequately evaluated by this test.

7.8 Conclusions and Recommendations

Conclusions: The fire safety record of Coast Guard-regulated vessels is good, relative to similar statistics in land transport or buildings. The fire safety regulations and specifications of polymeric materials in Coast Guard- and Navy-regulated vessels is much more stringent and closely controlled than similar regulations in land transportation or buildings. The absence of flammability requirements for hull materials used in naval ships could lead to the uncontrolled use of polymers in the future. *Recommendations:* It is recommended that flammability regulations be set on naval ship hull materials.

Conclusion: The fire safety specifications regulating polyurethane foams in both commercial and naval ships are inadequate. *Recommendation:* It is recommended that the fire safety regulations controlling the use of polymeric foams in both naval and commercial ships be reevaluated and strict standards be set that rigorously control the design use of polyurethane foam.

Conclusion: The large amounts of sensitive electronic equipment on naval combat ships and their need to maintain full operational capabilities at all times make these vessels particularly sensitive to the corrosive gases produced by fires in most of the cable insulation currently in use. *Recommendation:* The current research program to develop fire retardant electrical insulation which will not produce corrosive gases when burned should be expanded, and some interim fire protection standards that take account of corrosive combustion products should be promulgated.

Conclusion: The current fuel load calculation, based on the heat of combustion of wood, used to estimate the total fuel load in compartments, is misleading because it makes no provision for the much higher heats of combustion of polymers. *Recommendation:* The present technique for calculating the fuel load in ship compartments should be modified to make allowances for the higher (relative to wood) heats of combustion of most common synthetic polymers.

Conclusion: The flammability and smoke specifications controlling the use of polymers in shipboard furnishings currently allow the use of many highly flammable and smoke-generating polymers. *Recommendation:* The flammability and smoke specifications controlling the use of polymers in shipboard furnishings should be reevaluated, preferably using the scenario analysis technique; more appropriate specifications should also be set.

Conclusion: The flammability regulations governing the use of polymers in insulation in small communication wire are probably unrealistic and inadequate because tests are run on individual wires, while in actual use, wires are almost invariably combined in large cables or cableways. *Recommendation:* The flammability regulations governing the use of polymers as insulation on small communication wire should be reevaluated and reestablished to better correlate with actual conditions.

Conclusion: The present Coast Guard regulations allowing the use of unprotected polymeric foams in the cargo spaces of commercial shipping is a serious fire hazard. *Recommendation:* The flammability regulations allowing the use of unprotected polymeric foams in cargo spaces of commercial vessels should be reevaluated, using the scenario analysis technique.

Conclusion: The use of fire resistant compartment penetrations is an effective method of reducing the fire spread along cableways. *Recommendation:* Effective fire resistant breaks for cableway penetrations in compartment bulkheads should be required on all naval and commercial ships.

Conclusion: The flammability specifications controlling the use of polymeric foams on the bulkheads and water piping of Navy ships may be inadequate because of the flammability test chosen. *Recommendation:* The flammability test specifications regulating the use of polymeric foams in the engineering and cargo spaces of naval vessels should be reevaluated, and toxicity specifications for pyrolytic combustion gases from these materials should be set.

Conclusion: The flammability and smoke specifications set in MIL-STD-1623 for light diffusers in naval ships can allow the use of some highly flammable polymeric materials. *Recommendation:* The MIL-STD-1623 flammability and smoke specifications on light diffusers should be reevaluated in view of presently available state-of-the-art materials and new specifications should be set.

Conclusion: The present naval research programs to develop new fire retardant elastomers, cable coatings, and cushioning materials with less corrosive and toxic combustion products is addressing one of the most serious and current fire hazards on naval vessels. *Recommendation:* The Navy research programs to develop new fire retardant elastomers, cable coatings and cushioning materials should be continued and expanded.

Conclusion: As is more fully demonstrated in Chapter 3, scenario analysis is a powerful tool in materials selection, design, criteria, validation of test methods, the promulgation of specifications, and the choice of research and development objectives. *Recommendation:* Scenario analysis should be a required part of the design and construction regulations for all new ship construction.

Conclusion: The absence of toxicity standards of combustion gases in Navy regulations on materials leads to serious safety hazards on ships. *Recommendation:* The promulgation of toxicity standards for combustion gases from polymeric materials should be set to diminish the hazard of the use of these materials on naval vessels.

Conclusion: The present flammability standards based on ASTM E-84 used for controlling materials on Navy ships may be inadequate. *Recommendation:* The use of ASTM E-84 as a primary flammability test controlling materials used on Navy ships should be reevaluated.

REFERENCES

Anon, Plastics, pp. 93—95, March 1960.

Emmons, H. W., "Fire & Fire Protection," Scientific American, 231 (1) 21—7 (1974).

Hindersinn, R. R., Encyclopedia of Polymer Science and Technology, Wiley Interscience, New York: 1962—1972.

Hunter, L., Steamboats on the Western Rivers, Harvard University Press, Cambridge, Mass., pp. 272 ff and Appendix (1949).

Intergovernmental Maritime Consultative Organization, International Conference on Safety of Fishing Vessels, 1977. SEV/CONF/21, 31 March 1977.

McDaniel, D., "Noncombustibility: A Marine View," in Ignition, Heat Release, and Noncombustibility of Materials, ASTM STP 502, American Society for Testing and Materials, Philadelphia: pp. 24—34 incl., (1972).

National Fire Protection Association, "Fire Protection Handbook (14th Ed.)," Boston: Sec. 13—11 ff. (1976).

Maritime Administration, U.S. Department of Commerce, the National Shipbuilding Research Program, Plastics in Shipbuilding, in cooperation with Todd Shipyard Corporation, Washington, D.C., August 1977.

Offshore News, 5 June 1974.

Sheehan, D. F., "Progressive Fire Protection," SNAME Vol. 8 (1972).

SNAME T&R Bulletin 2—21, July 1974.

U.S. Coast Guard Document CG-129, Proceedings of the Marine Safety Council, March 1977 and March 1978.

1929 Safety of Life at Sea Convention, CG-242, U.S. Coast Guard Document dated June 1, 1951.

U.S. Delegation Report of the Secretary of State, dated April 13, 1948, 1948 Safety of Life at Sea Conference. U.S. Coast Guard Document CG-242 dated June 1, 1951.

U.S. Coast Guard Publication CG-329, Fire Fighting Manual for Tank Vessels, dated January 1974.

Secretary of State letter to President dated February 16, 1968, 90th Congress 1st Session, U.S. Senate Document, Executive C, dated February 28, 1968.

APPENDIX A — SOLAS CONVENTIONS

A1.1 Background

A1.1.1 National

Shipboard structural fire protection is one element of an overall "Life Safety System" incoproated into the current vessel regulations. The development of this "Life Safety System" began in the early Twentieth Century, when the SS TITANIC sank on April 14, 1912. The heavy loss of life was a primary cause for the calling of an international conference for the safety of life on the high seas. In 1914, the first International Conference on Safety of Life at Sea was held in London. The recommendations of the Conference concerned vessel subdivision and minimum requirements for lifesaving devices, with no mention of structural fire protection. Because of the First World War, the provisions of this conference were never fully implemented.

In 1929, a second conference promoting the Safety of Life at Sea was held in London. The purpose of this conference was to continue the development of international standards for the construction and arrangement of passenger vessels that was begun in 1914. On May 31, 1929, the "Convention for the Safety of Life at Sea" was signed. Regulation XVI of the Convention addressed structural fire protection, which required the fitting of fire-resisting bulkheads above the weather deck. These bulkheads were constructed of metal or other fire-resisting materials effective to prevent for 1 hour, under the conditions for which the bulkheads are to be fitted in the ship, the spread of fire generating a temperature of $1500°F$ $(816°C)$ at the bulkhead. Seven years passed before the Convention was ratified by the United States. Impetus towards the ratification of this document and the consequent development of shipboard structural fire protection measures was supplied in 1934 when the U.S. flag passenger vessel Morro Castle burned off the coast of New Jersey, causing the death of 124 persons. The public reaction to this was sufficient to cause the creation of a special subcommittee of the United States Senate Committee on Commerce to investigate the SS Morro Castle fire and to develop recommendations for "Life Safety" standards aboard U.S. vessels. The subcommittee was divided into groups, assigned to deal separately with the various elements affecting life safety at sea. The investigation of fire protection measures was assigned to the Subcommittee on Fireproofing and Fire Prevention under the leadership of George G. Sharp. In its report, the subcommittee noted, "The first problem confronting the committee was the question as to what general method of fire control might be the most practical combination of effectiveness and simplicity. Since past experience demonstrated the vulnerability of complex automatic and manually controlled systems (detection and extinction, widely spaced fire doors, etc.), it was agreed that, if possible and economically practicable, the most foolproof solution to the problem would be a construction that would confine any fire to the enclosure in which it originated." The 1929 SOLAS Convention required "fire resisting bulkheads," however, a precise definition or standard test for "fire resisting bulkheads" was not included in those regulations.

To develop an adequate and comprehensive definition of "fire resisting bulk-heads," the subcommittee decided to conduct a series of full-scale shipboard tests to evaluate several different methods of construction. A test ship, the SS Nantasket, was procured from the Reserve Fleet on the James River, and in mid-1936, numerous tests proved the effectiveness of one type construction which made use of steel and asbestos composition panels. This construction technique was recommended by the Marine Section of the National Fire Protection Association and involved two types of "fire resistive" bulkheads ("Class A-1" and "Class B"). The "Class A-1" bulkheads were intended for use as fire screens or main vertical zone bulkheads. They were made of steel and lined with insulating material to maintain structural integrity and prevent the spread of fire (on the unexposed side) when exposed to a standard fire test for one hour. The "Class B" bulkheads were intended for use in forming stateroom boundaries. They consisted of incombustible materials effective to maintain structural integrity and prevent the spread of fire (on the unexposed side) when exposed to a standard fire test for 30 minutes. The standard fire test recommended by the Marine Section of the N.F.P.A. was the laboratory fire endurance rating test used by the National Bureau of Standards which had been adopted as a standard test method in 1918 (ASTM E-119).

During the SS Nantasket tests, data was recorded to compare the temperatures in the test room to those generated in the standard laboratory test furnace. Initially, the SS Nantasket tests were conducted using clothing and furnishings as a fuel source. This proved to be unsatisfactory, since it produced very poor combustion. Cord wood was substituted as a satisfactory fuel source for the remainder of the tests. To approximate the energy content of the clothing and furnishings, a fuel load of 5 lbs/ft^2 was used. With this configuration, fires equivalent to the standard laboratory test of 15 and 30 minutes were achieved.

Based upon the test results, the Subcommitee reported to Congress, "It would be impossible to fireproof a modern passenger ship by the methods used ashore." During the SS Nantasket testing, it was determined that certain materials commonly used for building construction "gave off such quantities of fumes that it was found impossible to approach even a minor fire to extinguish it. During the course of experiments, a form of construction was developed in which combustible material was eliminated to such an extent that combustion cannot be sustained by any part of the ship's structure." This form of construction used steel and asbestos composition test panels, which proved to be far superior to the fire retardant wood and fire retardant wood panels (faced with steel) in the test series.

As a result of the recommendations presented by the Subcommittee in Chapter IV of Senate Report No. 184, the United States Congress ratified the 1929 Convention for the Safety of Life at Sea, and amended the United States Code to require U.S. flag vessels to employ the use of "fire retardant material in their construction so far as is reasonable and practicable." Although it was not clearly defined in the United States Code, the type of construction that proved successful in the

"Nantasket tests," was intended. This construction consisting mainly of steel and asbestos composition panels could be considered incombustible by most test methods, and provided little additional fuel loading.

Under the authority of U.S.C. 369, the Secretary of Commerce promulgated Order No. 42 on July 17, 1940. Order No. 42 created Part 144 of Title 46 of the Code of Federal Regulations or Subchapter M. Paragraph 144.4(a) of Subchapter M required that interior boundaries be constructed of Class A-1 (A or B) fire retardant materials. Class A-1 bulkheads were required to be steel, lined or insulated with incombustible materials to prevent the average temperature on the unexposed side of the test bulkhead from rising more than 250°F (139°C) or any single-point temperature from rising more than 325°F (169°C) in 1 hour when subjected to the standard fire test. Class A bulkheads were required to be steel and to withstand the standard fire test for 1 hour, with no temperature rise limitations. Class B bulkheads were required to be incombustible materials capable of withstanding the standard fire test for 30 minutes and to also be capable of preventing the aforementioned temperature rise limitations for 15 minutes. Again, the terms "fire retardant" and "incombustible" were used without precise definition. Unfortunately, there were materials that could be considered fire retardant and which could pass the standard fire test, but did not have the equivalent combustibility properties as steel or asbestos. Because of the lack of a specific test method, certain materials could be approved which had the potential to greatly contribute to the fuel load of a protected space. It was not until the end of World War II that a specific test was developed to classify materials as incombustible. In 1949, the Coast Guard adopted Standard 46 CFR 164.009 (incombustible materials for merchant vessels) based upon research conducted at the National Bureau of Standards by N.P. Setchkin and S.H. Ingberg.

The question of fuel loading had also been in question for several years. During World War II, the need for lighter weight ship structures had brought about the use of aluminum bulkheads, and after the war aluminum bulkheads were proposed for use aboard passenger vessels. It was agreed that aluminum bulkheads would be an acceptable substitute for the asbestos panels, even though they do not withstand the standard fire test. The basis for this argument was that aluminum has a very high thermal conductivity that tends to dissipate heat rapidly, and secondly, that advocates of aluminum felt that the intensity of the fires in the "Nantasket tests" was due to the cordwood fuel source. They maintained that the typical contents of a stateroom would not constitute a fuel load sufficient to cause melting of the bulkheads. Therefore, in 1947 a full scale "stateroom burnout test" was conducted in conjunction with Gibbs & Cox, Inc. and the National Bureau of Standards. The stateroom test was conducted in a mock-up stateroom using typical furnishings and the personal belongings of three passengers as a fuel source. It was estimated that the fuel source was equivalent to 4,300,000 Btu's. This test verified the results of the "Nantasket tests," and showed that a fire involving only typical stateroom furnishings is capable of generating the same temperatures as the standard fire test laboratory furnace. The stateroom test also showed that uninsulated aluminum bulkheads could not provide the same degree of fire endurance as asbestos composition panels.

The new marine technology created during World War II was the cause for a third Interntional Conference on the Safety of Life at Sea, held in London in April 1948. Naturally, the United States proposed the incorporation of fire protection techniques as listed in Subchapter M of Title 46. Because of the materials employed and because certain nations felt that active fire protection systems were equivalent to passive fire protection, three alternate methods of shipboard fire protection were listed in the 1948 Convention: Method I (Subchapter M), was the technique proposed by the United States; Method II, proposed by the United Kingdom, advocated the use of sprinklers with no restricton on the combustibility or fire endurance of compartment bulkheads; and Method III, proposed by France, made use of a limited amount of fire-resisting bulkheads in conjunction with a fire detection system. The 1948 Convention came into effect in the United States on Nov. 19, 1952. To implement the provisions of this document, and to revise the passenger vessel inspection regulations into one subchapter, the Coast Guard withdrew Part 144 and created a new Part 70 or Subchapter H in Title 46. The regulations written for this new subchapter are basically those in effect today.

It is interesting to note the changes made regarding bulkhead fire endurance ratings in the new Subchapter. The old Class A-1 bulkheads are now A-60, Class A was changed to A-0, and the Class B bulkheads were now B-15. Two new categories of bulkheads were created: A-30 Class bulkheads were an intermediate A Class bulkhead and B-0 Class bulkheads were created because the former B-Class bulkhead panels had an inherent 15-minute fire endurance rating; however, unless certain connectors or "H-posts" were used, a heat transfer through the connectors occurred. It was felt that if these bulkheads were installed next to spaces with very low fuel loads, such as toilet spaces, a B-0 Class rating would be acceptable.

A1.1.2 International

The Intergovernmental Maritime Consultative Organization is a specialized agency of the United Nations concerned with establishment of safety and pollution prevention standards for vessels engaged in international trade. This organization had its genesis in 1948 when, "The Transport and Communications Commission of the Economic and Social Council of the United Nations recommended that the United Nations sponsor the creation of an international maritime body, and, accordingly the United Nations Maritime Conference was held at Geneva Feb. 19—March 6, 1948, resulting in a draft convention for an Intergovernmental Maritime Consultative Organization (IMCO)." (U.S. Delegation Report, 1951). Shortly after this meeting, an International Conference on the Safety of Life at Sea (SOLAS) was held in London, beginning April 23, 1948. This conference noted the tentative establishment of a mechanism (IMCO) to upgrade on a continual basis the various technical treaties which previously had been changed on an infrequent basis.

The organization came into being on Jan. 6, 1959. It is located in London, and continues to serve a vital role in the coordination and conduct of safety matters related to the maritime world.

Prior to the establishment of IMCO, conferences were called at infrequent intervals primarily in response to disasters. For example the 1914 and the 1929 Conferences for the Safety of Life at Sea were called primarily in response to the SS TITANIC disaster. As such they dealt primarily with standards for subdivision, minimum life saving appliances, and the use of radio. Also established was the International Ice Patrol and the use of fixed routes on the North Atlantic run was recommended.

There have been three subsequent conferences concerning Safety of Life at Sea (SOLAS), and each has progressively upgraded the requirements for fire protection. It is necessary, for the purpose of this report, to examine the 1948, 1960, and 1974 SOLAS Conventions, since new vessels are currently being built to higher standards, while many existing vessels were built to older minimum standards as specified by the earlier conventions.

The following discussions will primarily address the accommodation and service sections of a ship (normally the crew and passenger berthing areas). Passive fire protection (bulkheads and decks), as opposed to active fire protection (fire pumps and fire mains), will be addressed.

A1.1.2.1 1929 SOLAS Convention

The primary passive fire protection requirement contained in the 1929 Convention was in Chapter II, Regulation XVI, entitled "Fire Resisting Bulkheads." The text is worth repeating, as it was a significant first step.

1. "Ships shall be fitted above the bulkhead deck with fire-resisting bulkheads which shall be continuous from side to side of the ship and arranged to the satisfaction of the Administration.

2. "They shall be constructed of metal or other fire-resisting material effective to prevent for one hour, under the conditions for which the bulkheads are to be fitted in the ship, the spread of fire generating a temperature of 1500°F (815°C) at the bulkhead.

3. "Steps and recesses and the means for closing all openings in these bulkheads shall be fire-resisting and flame tight.

4. "The mean distance between any two adjacent fire resisting bulkheads in any superstructure shall in general not exceed 132 feet (40 meters)." (1929 Safety of Life at Sea Convention, U.S. Coast Guard, 1951).

This regulation only applied to passenger vessels on international voyages.

Polymeric usage at that time was essentially limited to natural polymers, particularly wood. Polymer usage was virtually uncontrolled, with no limit to the amount, type, or application.

The passenger vessels of the period were opulent examples of post-Victorian elegance. The trade of marine joiner work (carpentry) had reached new heights, due to advances in power machinery. The only control exercised over the use of combustibles was the physical limitation of the vessel itself. While steel was beginning to make greater and greater inroads into shipbuilding, it was primarily utilized as a hull material.

A1.1.2.2 1948 SOLAS Convention

This conference convened at the urging of major maritime nations and accomplished what might be considered a quantum leap with respect to more rigorous requirements for passenger vessels. It should be noted that while no specific passive fire protection methods were detailed for cargo vessels, requirements for active fire protection systems were beginning to take form. During the period between the the coming into force of the 1929 SOLAS Convention, and the call for a post-war conference, the major maritime nations of the world evolved methodologies for the protection of passenger ships against fire. McDaniel details the rationale developed by the United States in its regulatory approach toward fire safety of passenger vessels. (McDaniel, D.E., 1972).

During the conference each of the proponents of the three methods battled for the emergence of their national method. The report to the Secretary of State by The United States Delegation to the 1948 Convention (1951) stated:

"The new convention includes regulations for fire protection in accommodation and service spaces on passenger vessels; the 1929 Convention contained only rudimentary provisions on this matter; the 1948 Convention recognizes the following alternatives of acceptable methods of fire protection:

1. "Method I (United States practice and proposal), regulates on the basis that all internal divisional bulkheading be made of essentially incombustible material. This represented the most severe limitation with respect to control of combustible polymers. The hulls, decks, structural bulkheads, superstructure, and deck house constructions are required to be of steel. Internal divisional bulkheading as well as insulation was required to be non-combustible as determined by test. Limited amounts of combustible veneers could be utilized for interior finish: however, these are specified to be of the low-flame-spread type.

2. "Method II (U.K. practice and proposal), is based on the adoption of an automatic sprinkler and fire alarm system generally, with no restriction on the type of internal divisional bulkheading. Except for those bulkheads and divisions required to be of steel or "A" Class construction, no restriction on the amount of combustibles is made. This approach was rationalized by the premise that automatic sprinkler protection would be utilized as protection against any fire. Internal divisions are not required to have any fire integrity.

3. "Method III (French practice and proposal), regulates on a system of subdivision forming a network of fire-retarding bulkheads enclosing limited areas,

175

together with the installation of a fire-detection system. This method essentially places no control over limitation on the usage of polymers, however, the vessel is further subdivided by requiring "B" Class non-combustible bulkheads which form a continuous network not exceeding 1,300 ft^2.

The three methods require the same basic fire-zone subdivision, enclosed fire-escape stairways and enclosures, and protection to prevent drafts and the spreading of fire to vertical trunks for elevators, electric cables, etc. The decision to adopt three alternative methods was necessitated not only by a desire to permit the systems of defense against fire, which the respective countries considered equal, but also by the practical consideration that the materials required by the United States system were not presently available internationally in sufficient quantities. The incorporation of regulations detailing definite requirements for defense against fire is considered to be one of the outstanding accomplishments of the convention."

The degree of polymer control varied with each method. A detailed analysis will be given in a later section.

While the ratification of the 1948 SOLAS Convention by the required number of countries was a major step forward, the United States was concerned that the three methods of construction which were afforded equal stature under the 1948 SOLAS Convention were in fact not equal. The basic question was whether the fitting of active fire protection systems provided the degree of protection afforded by inherent fire safety brought about by the limiting of combustible materials. This debate continued until the next International Conference on Safety of Life at Sea in 1960.

This conference was again in response to a highly visible maritime disaster; the collision of the SS ANDREA DORIA and SS STOCKHOLM. The calling of an international conference normally is in response to specific problems; however, it normally provides an opportunity to upgrade the broad spectrum of requirements.

A1.1.2.3 1960 SOLAS Convention

The 1960 SOLAS Convention, while essentially maintaining the three primary methods of achieving fire protection, significantly tightened the minimum requirements for cargo vessels. Additionally, requirements were formulated for small passenger ships. As part of the Convention numerous recommendations were made concerning fire protection.

In the mid-1960s, a series of serious passenger vessel fires occurred, which again raised the question of the adequacy of fire safety standards of vessels built to the 1929 and 1948 SOLAS Convention. The following is a portion of the letter of transmittal from the U.S. Secretary of State to the President concerning the actions taken internationally to upgrade passenger vessel safety.

"After the disastrous Yarmouth Castle fire in November 1965, the United States proposed a reconsideration of the fire safety rules by IMCO on an urgent basis. First attention was given to the most urgent problem, the rules for existing passenger ships. Amendments to the Convention to upgrade the standards for existing ships were approved by the IMCO Assembly in November 1966 and have subsequently been accepted by the United States. In the meantime IMCO continued work on new regulations for passenger ships to be constructed in the future. Agreement was reached on construction methods based on the maximum use of incombustible materials and a series of new regulations was approved and recommended by the IMCO Maritime Safety Committee in March 1967 and adopted by the Assembly in October 1967. When these amendments come into effect the task of upgrading fire safety standards in the aftermath of the Yarmouth Castle disaster will be completed.

All of these amendments were adopted by the IMCO Assembly either unanimously or by overwhelming majorities. There were a few countries which criticized the new passenger ship fire safety standards essentially as too severe in requiring the use of incombustible materials, but the very great majority of members supported the maximum use of incombustibles, as reflected in the amendments and advocated by the United States." (Secretary of State Letter, 1968).

The 1966 Fire Safety Amendments were unilaterally enforced by the United States by Public Law 89-777. The essence of this law was the requirement that any vessel which embarked more than 50 U.S. passengers from U.S. ports on an overnight voyage would be required, as a minimum, to comply with the 1966 Fire Safety Amendments.

The basic requirements that are detailed in the various conventions require that either: (a) there is universal understanding with respect to intent; (b) agreement on a basic test methodology to assure consistent performance; or (c) the matter is resolved by allowing the national administration to determine the basic requirements. Emmons (1974) found the latter approach questionable at best even when attempting to define the same condition. In an attempt to bring about conformance with the stated objectives of fire safety testing the following test methods were developed by the IMCO forum and formalized into recommended test procedures:

1. IMCO Resolution A.270 (VIII) "Recommendation on Test Methods for Qualifying Marine Construction Materials as Non-Combustible."
2. IMCO Resolution A.163 (ESIV) and A.215 (VII) "Recommendation for FIRE Test Procedures for "A" and "B" Class Division."

Currently the fire Test Working Group, of the IMCO Subcommittee on Fire Protection, is considering development of the following methodologies and protocols:

1. Development of an international standard for fabric flammability to meet the requirements of SOLAS 1974.
2. Investigation and formulation of a standard for IMCO/ISO flame spread test apparatus.
3. Completion of an international round-robin test program for carpet and floor flammability.

Passenger vessel requirements were not significantly changed from the 1948 Convention, however, cargo vessels of 4,000 gross tons and upward were required to be constructed of steel. Corridor bulkheads were required to be manufactured from steel or "B" Class materials. This essentially provides a fire boundary rather than increasing the internal fire safety.

A1.1.2.4 1974 SOLAS Convention

When this convention comes into force a major international legal instrument will severely restrict the use of polymeric materials aboard two major classes of vessels — passenger vessels and tank vessels. Passenger vessels will essentially require a unified approach which the Maritime Safety Committee of IMCO described as "maximum use of non-combustible materials and the appropriate use sprinklers and detecting systems." This approach permits the best of both worlds with respect to active and passive fire protection systems. Non-combustible materials for basic divisional structure coupled with detection or sprinkler systems provide a multi-level arsenal of protection.

Tank vessels will be required essentially to mirror the requirements for use of non-combustible materials. However, the stringent insulation values required between adjacent compartments on passenger ships will be significantly reduced.

A1.1.2.5 1977 Fishing Vessel Conference

In March, 1977, the first major international conference concerning safety of fishing vessels was held in Spain. While fire safety was not the primary area of emphasis there was a recognized concern for increased polymer usage being experienced in the construction of fishing vessels. One of the final recommendations was entitled "Guidance Concerning the Use of Certain Plastic Materials."

"In considering the problem concerning the use of certain plastic materials, particularly in accommodation and service spaces and control stations, the Administration should note that such materials are flammable and may produce excessive amounts of smoke and other toxic products under fire conditions." (Intergovernmental Maritime Consultative Organization, 1977).

The usage of polymers aboard vessels which are required to comply with requirements laid down by international convention has gone through a cyclical change. The trend within international deliberation is towards the controlled usage of polymers. This trend should have a dramatic positive impact on the overall fire safety of vessels in international trade.

As can be seen by these conventions, progress in the standard-making process is often painstakingly slow; however, the rewards are fulfilling.

APPENDIX B

Department of Transportation U.S. Coast Guard

Equipment Lists

Items Approved or Accepted under Marine Inspection and Navigation Laws

CG-190

U.S. Coast Guard, Washington, D.C., 20590

NOTE: This page is inserted to direct the reader to the source of this information.

APPENDIX C

Navigation and Vessel Inspection Circular — NVC 10-63

Sketches of Bulkhead and Deck Construction

Department of Transportation
U.S. Coast Guard, Washington, D.C. 20590

NOTE: This page is inserted to direct the reader to the source of this information

APPENDIX D

FIRE LOADING CALCULATIONS

The following information applies to fire load calculations to determine the insulation thickness for aluminum structures, thereby ensuring structural integrity, and to limit the temperature rise of the unexposed surface for a predetermined period of time.

The criteria apply only to passenger and cargo ships. For tank vessels, the fire exposure situation, external and internal, is different and therefore these guidelines do not apply.

Nomenclature

F   effective insulating value of insulation protecting the aluminum core
F   effective insulating value of insulation on exposed face of bulkhead
F   total insulating value of bulkhead insulation
F   effective insulating value of insulation on unexposed face of bulkhead
P   bulkhead panel thickness approved under USCG specification 46CFR 164.008
S   structural insulation thickness approved under USCG specification 46CFR 164.007

D1.1 Fire Loading Calculations

To calculate fire loading, it is first necessary to determine which compartments are required to be separated by effective fire boundaries. Each situation is unique in this regard, but several general guidelines may be applied:

1. Normally, spaces of unlike character, (e.g., a stateroom and a corridor) must be separated by effective fire boundaries. As exceptions to this, private toilet spaces may be considered part of the cabin they serve, and spaces directly associated with another, (e.g., a pantry annexed to a galley) may be considered together.

2. On passenger ships, like compartments, (e.g., adjacent staterooms) must normally be separated by effective fire boundaries.

3. On cargo ships, several like compartments, (e.g. adjacent staterooms) may be included in areas separated by effective fire boundaries, providing no such bounded area exceeds 550 ft^2 (50m^2).

Next, it is necessary to calculate the total deck area in square feet of each compartment bounded by effective fire barriers.

Finally, one must calculate the total weight of combustibles[1] within each "fire" compartment. Validity of the fire loading approach depends entirely upon comprehensive and detailed estimates of combustible contents within a compartment. Calculations should include all combustible portions of:

---

[1] Heat of combustion varies with the material but, for the usual range of shipboard combustibles, assuming an average heat of combustion equivalent to wood is acceptable.

fixed items:	wall and ceiling linings, floor covering, electrical wiring, light diffusers, mouldings, etc.
furnishings:	furniture, mattresses, curtains, decorations, etc.
contents:	clothing and personal effects (staterooms), life jackets, cargo, stores, etc.

The weight of combustibles is then divided by the floor area of the compartment to give the fire loading of that compartment (lb/ft^2).

It is unlikely, despite conscientious efforts, that a fire loading analysis during the design phase of a vessel will include all combustibles which are actually installed. To allow for this, and to account for minor changes of occupancy during the life of a vessel, a safety factor is needed. These guidelines include such a safety factor in equating combustible loading to fire endurance. For example, 5.0 lb/ft^2 combustible loading is equivalent to a 30-minute fire. These guidelines permit a maximum loading of only 4.49 lb/ft^2 in a space whose boundaries are insulated for 30 minutes, a 10 percent safety factor. This approach actually provides a sliding safety factor, one which is higher at lower calculated fire loadings. Because of this approach, an additional safety factor need not be added to the calculated fire loading.

For guidance, estimated fire loadings for "typical" shipboard spaces are listed in Table D-1. Certain types of spaces, such as machinery spaces, cargo spaces, and storerooms for combustible material should always be assumed to have the fire loading shown.[2] Other spaces, indicated on the list, may have combustible loadings which differ from those shown.

Calculated fire loadings must be approved by the Coast Guard. It may be presumed that the Coast Guard would look askance at any design purporting to have minimum furnishings inconsistent with modern comfort standards and/or the number of persons allowable on board. In other words, fire loadings significantly lower than those shown in Table D-1 should be justifiable in considerable detail.

Since insulation is based upon presumed fire loading in each "fire" compartment, a record of calculated and allowable fire loadings should follow the vessel throughout its life. One method of achieving this is to indicate relevant information on the fire control plans required to be carried aboard the vessel. The Coast Guard should be consulted regarding the method proposed. It may be expected that, additional to the above measures, the vessel safety certificate will be endorsed to indicate that the installed fire protection is based upon calculated fire loadings recorded elsewhere.

D1.1.1 Duration of Protection Required

Calculation of fire loading may reveal minor variations among similar spaces. To

---

[2] Some types of specialized service vessels such as hydrofoils and hovercraft which are designed to have rapid debarkation using lifesaving equipment and which have automatic fire detection and fixed extinguishing equipment in machiner spaces, etc., may have such spaces insulated for only 30 minutes, subject to special Coast Guard approval.

Table D-1. Typical Fire Loading of Various Spaces

(*indicates calculated fire loading for individual compartments may not differ from typical values shown. Where no * appears, listed fire loading may be modified by calculation.)

Control Spaces

Wheelhouse: Chartroom	1.5 lb/ft^2
Fire Control Stations	1.5 lb/ft^2

Escape Routes

Corridors	1.5 lb/ft^2
Stairway enclosures	1.0 lb/ft^2

Accommodation Spaces

(Allowance to be made for personal effects as follows: staterooms — 1.5 lb/ft^2, public spaces — 0.15 lb/ft^2; these figures included in the following values)

Staterooms

Fire resistant furnishings	3.0 lb/ft^2
Combustible furnishings	5.0 lb/ft^2

Public Spaces

Fire resistant furnishings

Lounges, Restaurants, etc.	3.0 lb/ft^2
Ferry Vessels	1.5 lb/ft^2
Combustible Furnishings	5.0 lb/ft^2
Sanitary Spaces (not part of stateroom)	0.4 lb/ft^2

Service Spaces

Galleys	10.0 lb/ft^2
Pantries (no heating appliances)	4.0 lb/ft^2
Food concession (ferry vessels no combustible stowage)	1.5 lb/ft^2
Workshops	10.0 lb/ft^2 *

Storerooms

Combustible	10.0 lb/ft^2 *
Cleaning Gear only	3.0 lb/ft^2

Laundries

Ship's Laundry	10.0 lb/ft^2 *
Private Use	1.5 lb/ft^2

Main Machinery and Cargo Spaces	10.0 lb/ft^2
Auxiliary Machinery Spaces	5.0 lb/ft^2
Tanks & Voids	0.0 lb/ft^2

---

avoid proliferation of construction criteria, such spaces should generally be placed in a single category and assumed to have identical fire loadings. For example, all staterooms should be calculated on the same basis. Additionally, only five standard time periods should be employed as follows:

Table D-2. Duration of Protection for Various Fire Loadings

Fire Loading (lb/ft^2)	Duration of Protection (min)
less than 0.5	0 (no insulation)
0.5 to 1.99	15
2.0 to 4.49	30
4.5 to 6.99	45
7.0 and above	60

Except where they form a portion of stairway enclosures, the vessel's exterior, or boundaries of control spaces, bulkheads requiring 30 minutes or less integrity may be Coast Guard approved bulkhead panels (CG Specification 46 CFR 164.008). In all other cases, including the exceptions above, the fire division (bulkhead or deck) must have a metal core.

D1.2 Insulation Required for Specific Fire Loadings

By use of calculated "F" values developed in preceding sections, it is a simple matter to relate fire loadings to the required "F" value of the boundaries. This is done in Table D-3. The required "F" value must, of course, be determined for each side of the fire division.

Table D-3. "Fc" Value Required for Bulkheads and Decks

Fire Loading (lb/ft^2)	Duration of Protection (min.)	"Fc" value
Less than 0.5	0 (no insulation)	0
0.5 to 1.99	15	0.25 "S"
2.0 to 4.49	30	0.45 "S"
4.5 to 6.99	45	0.61 "S"
7.0 and above	60	0.72 "S"

D1.1.3 Limited Unexposed Surface Temperature Rise

The foregoing analysis assumed it was only necessary to determine the thickness and arrangement of insulation necessary to limit temperature rise of the aluminum to 200°C (360°F) or less. In most instances this would be adequate. If, however, it is necessary to limit the temperature rise on the unexposed face of the bulkhead, such as when combustibles may be stored in contact with the unexposed face, the total insulating value of the bulkhead, F_t, must be adequate for the required insulating period. The F_c required for various insulating periods is given in Table D-4.

Table D-4. Minimum F_c and F_t Values for Aluminum Divisions

		Limitation of Core Rise to 200°C (360°F) for				
		0 Minutes	15 Minutes	30 Minutes	45 Minutes	60 Minutes
Limitation of Unexposed Face Rise to 139°C (250°F) for	0 Minutes	$F_c = 0$ $F_t = 0$	$F_c = 0.25$ "S" $F_t = 0$	$F_c = 0.45$ "S" $F_t = 0$	$F_c = 0.61$ "S" $F_t = 0$	$F_c = 0.72$ "S" $F_t = 0$
	15 Minutes	X	$F_c = 0.25$ "S" $F_t = 0.50$ "S"	$F_c = 0.45$ "S" $F_t = 0.50$ "S"	$F_c = 0.61$ "S" $F_t = 0.50$ "S"	$F_c = 0.72$ "S" $F_t = 0.50$ "S"
	30 Minutes	X	X	$F_c = 0.45$ "S" $F_t = 0.70$ "S"	$F_c = 0.61$ "S" $F_t = 0.70$ "S"	$F_c = 0.72$ "S" $F_t = 0.70$ "S"
	45 Minutes	X	X	X	$F_c = 0.61$ "S" $F_t = 0.86$ "S"	$F_c = 0.72$ "S" $F_t = 0.86$ "S"
	60 Minutes	X	X	X	X	$F_c = 0.72$ "S" $F_t = 1.0$ "S"

X = These conditions are not applicable, since limitation of core temperature rise is for a shorter period than limitation of unexposed face temperature rise.

APPENDIX E

SHIPPING CONTAINER FIRE-TEST EXPERIMENT

(Factual Information Supplied by the U.S. Coast Guard)

E1.1 Background

The shipping container fire-test scenario (test series) was conducted to obtain credible data for three different types of standard shipping containers (steel, aluminum, and fiberglass reinforced plywood). Due to the lack of realistic experience in this area, it is virtually impossible to accurately predict the flame spread and course for this type of fire.

Since the advent of container shipping over two decades ago, there has been only one major casualty aboard an American flag container vessel. The collision of the SS C.V. Sea Witch with the SS Esso Brussels on June 2, 1973, and the resulting fire caused the complete destruction of all 285 containers, including deck cargo stowed on the Sea Witch. The lack of credible data for this type of occurrence was a setback that prompted the U.S. Coast Guard to sponsor and work hand-in-hand with an ad hoc advisory group comprised of fire protection and marine experts from government and industry.

The fire which destroyed the containers and cargo did not start within the container stack, but had spread from a pool fire of more than 1 million gallons of Nigerian crude oil surrounding the container vessel. The heat flux experienced by the containers from this exposure was well in excess of that normally produced in laboratory furnace tests when determining the fire endurance of materials.

Evaluation of the routine transport of intermodal shipping containers envisaged several fire scenarios among which are: (1) a likely fire incident may occur as a result of shifting cargo within a properly lashed container, and (2) a second likely fire incident may occur as the result of a flammable fluid leak, where a container stack is exposed to a limited external fire source. If either credible fire source, interior or exterior, were exposed to one container or to several stacked containers, an estimate of the potential flame spread to the remainder of the container load is necessary.

Tests evaluated the fire potential and flame spread of different material combinations used in the construction of standard 8-by-8 by 20-foot shipping containers (steel, aluminum, and fiberglass plastic reinforced plywood). For purposes of this publication, FRP containers are used singly and in combinations with aluminum and steel containers (see Tests 1 through 6 in Section E.1.2.5.1).

E.1.2 Procedure

E.1.2.1 Facility

The container fire scenario test series was conducted at the Coast Guard's Fire and Safety Test facility in Mobile, Ala. Because gantry cranes or other container lifting devices could not be practically erected, the fire scenario tests were not conducted aboard one of the Test Facility vessels. A test pad, measuring 3,600 ft$_2$, was constructed on Little Sand Island, adjacent to the SS Albert E. Watts. A simulated hatch cover, measuring 29 by 25 feet, was constructed on the test pad by welding steel plates together. Steel coamings were welded along the perimeter of the hatch cover to form a 10-inch-deep fuel pan for the pool fires. Two steel I-beams, with six inch flanges, were fitted with bottom stacking fittings and pad eyes, and centered in the fuel pan to act as a base for the container stack.

E.1.2.2 Fuel Source

1. *Internal* — Standard wood cribs weighing 28.5 to 31.5 pounds and constructed of forty-five pieces of white fir measuring 2 by 2 by 15 inches were used as a Class A fuel source. In Tests 1 through 3A, two wood cribs were stacked vertically over a 13-by 13-by 4-inch steel pan containing 2 gallons of naphtha. In Tests 1 through 3A, the fuel source was located at the center point of the container floor. Test 3A was conducted with the wood crib and naphtha pan located 5 inches from the starboard or rear corner of the aluminum container.

2. *External* — Aviation fuel (JP-5) was used as a fuel source for all external exposure tests. Several gallons of naphtha were used to prime the JP-5 for easier ignition.

E.1.2.3 Instrumentation

The Fire and Safety Test Facility instrumentation van was used for this fire test series. To facilitate the connection of all necessary wiring and electrical power supply circuits, the van was loaded on one of the Test Facility LCM's, which was then moored on the island adjacent to the test site.

For all of the interior fire scenario tests, internal temperatures, oxygen, carbon monoxide, and carbon dioxide levels were measured and recorded. On the external exposure fire scenario tests, only temperatures were recorded. The thermocouples and gas sensor tubes were routed into each container through small holes drilled in each container door. The openings were then sealed with a high-temperature caulking material. The thermocouples used were Type K, ungrounded (sheathed), 10-inch-diameter Inconel.

Oxygen concentrations were measured with a thermomagnetic oxygen analyzer with a range of 0 to 25 percent, 1.0 percent. Carbon monoxide concentrations were measured with a Luft type infrared analyzer with a range of 0 to 50 percent, + 1.0 percent. All instrumentation data was fed into an analog to digital converter whose output was recorded on both printed-paper-tape and paper-punch-tape. The punch-tape data was fed into a computer, which plotted the data as engineering units versus time.

E.1.2.4 Ignition Method

For the internal tests, sufficient lengths of fuse cord were run from the naptha pan at the base of the wood crib to the exterior of the container. The fuse cord was then ignited (manually) from a safe distance.

E.1.2.5 Containers

Generally, dry cargo intermodal shipping containers consist of two side panels, one end panel, doors, roof, and floor. These components are held together by the container frame components which consists of two bottom rails, two top rails, and two end frames. The frame components are usually high-tensile-strength steel or extruded aluminum alloy. Side- and end-panels consist of varying materials, based primarily upon the nature of the cargo to be carried. The containers evaluated in this series involved four basic types of panels — steel, fiberglass plastic reinforced plywood, exterior post aluminum, and exterior skin aluminum panels. Container doors, in most cases, are exactly the same as the container's side panels, or they may be a composite material referred to as Plymetal. Plymetal doors consist of a plywood core with aluminum or galvanized steel sheeting on both exposed surfaces. Container floors are constructed of oak or other hardwood floorboards supported by the frame crossmembers. The floorboards are generally connected to one another by either tongue and groove or ship lap constructions. Container roofs are constructed of materials similar to the container side panels.

All joints formed by the connection of a frame member to a panel are sealed with a weatherproof compound or gasket.

E.1.2.5.1 Container Materials and Construction Details

Containers 1 through 9 were used in fire Tests 1 through 9. Containers 4, 5, and 6 were used for interior fire Tests 1, 2, 3, and 3A. The containers used for exterior fire Tests 4 and 5 were the same units used in Tests 1 through 3A. The containers used for Test 5 were deliberately destroyed and were replaced with identical types for Test 6. The containers used for the tests are listed as follows by test number and container number and are described in Table E-1.

Test No.	Container No.
1	4
2	5
3	6
3A	4, 5, and 6
4	4, 5, and 6
5	4, 5, and 6
6	1–9

E.1.3 Interior Fire Scenario

The single factor capable of regulating combustion of fuels within a sealed container is the amount of available oxygen. Assuming no leakage, the maximum

188

Table E-1. Container Materials and Construction

<u>Container Nos. 1 and 7</u> - Veenema & Wiegers, Inc., Model M-20880 AP, 1966

- Side and end panels	- 3/4 inch plywood core sandwiched between 0.060 inch polyester fiberglass
- Top rails, bottom rails, or end frames	- 6061-T6 or 6062-T6 aluminum alloy
- Roof	- 5/8 inch plywood core sandwiched between 0.060 inch polyester fiberglass
- Floors	- oak edge grain, 1 1/8 inch laminated
- Lining	- 1/4 inch A.C. plywood, full height

<u>Container No. 2</u> - Stick "Sea Trailer"

- Side panels	- 0.051 inch aluminum sheet, extruded aluminum alloy posts on 24 inch centers

<u>Container No. 4</u> - SIC steel container, made in France

- 4 fork lift pockets

- 4 ride vents

- 1 1/4 inch floorboards

- SIC locking system

<u>Container No. 9</u> - Steel container "Tokyo Car."

- End panel	- 0.051 inch aluminum sheet, 5 posts on 12 inch centers
- End frames	- high tensile strength steel
- Roof	- 0.050 inch aluminum sheet, bows on 24 inch centers
- Floors	- 1 1/4 inch floorboards with 5 inch (10 USCG) high tensile strength steel crossmembers
- Doors	- Plymetal - 3/4 inch plywood core, 0.025 inch aluminum sheet on each side
- Lining	- 1/4 inch plywood, 48 inches high on sides, and full height on end panel

---

(Continued)

Table E-1. Container Materials and Construction (Continued)

<u>Container Nos. 3 and 8</u> - Fruehauf, Model KA2-20

- - Side and end panels   - 0.063 inch aluminum sheet

    1 3/10 inch posts on 18 inch centers

- - Top rails or bottom rails   - 6061-T6 aluminum alloy

- - End frames   - high tensile strength steel

- - Roof   - 0.063 inch 3003-H14 aluminum sheet

- - Floors   - 1 1/16 inch hardwood, tonque and qroove floorboards over I-beams on 12 inch centers

- - Doors   - 3/4 inch Plymetal

<u>Container No. 5</u> - Theurer, Model PC20-8

- - Side and end panels   - 3/4 inch plywood core sandwiched between 0.060 inch polyester fiberqlass

- - Front end frame   - aluminum alloy

- - Rear end Frame   - high tensile strength steel

- - Foof   - 5/8 inch plywood core sandwiched between 0.060 inch polyester fiberqlass

- - Doors   - Plymetal, type 8F2, 0.050 inch aluminum sheet exterior, 24 quaqe steel interior, 3/4 inch plywood core.

- - Upper side rails   - extruded aluminum alloy

<u>Container No. 6</u> - Fruehauf, Model KF2-20

- - Side panels   - 0.063 inch aluminum sheet, 4 by 1 3/16 inch (6061-T6) aluminum alloy posts on 24 inch centers

- - End panel   - high tensile strength steel frame, 0.063 inch aluminum sheet, for interior posts

- - Roof   - 0.063 inch 3003-H-14 aluminum sheet with 1 inch deep (6061-T6) I-beam bows on 24 inch centers

- - Doors   - 1 inch Plymetal consisting of exterior 0.040 inch aluminum sheet, 24 gaqe zinc plated steel sheet on interior

---

(Continued)

Table E-1. Container Materials and Construction (Continued)

- Floor   - 1 1/8 inch laminated hardwood
tonque and groove

Container No. 4 - SIC steel container, made in France

    - 4 fork lift pockets

    - 4 ride vents

    - 1 1/4 inch floorboards

    - SIC locking system

Container No. 9 - Steel container "Tokyo Car."

(Concluded)

quantity of air in one of the shipping containers is 1,280 cubic feet. A loaded container will naturally have less than 1,280 cubic feet of air available for combustion. Therefore, to simulate a maximum or worst case test situation, it was decided to test the containers filled with as little cargo as possible. Because of the variety of cargo that is stowed in containers, it was also decided to utilize both Class A and Class B fuels.

Calculations using the combustion engineering formula indicated that the wood cribs, used as a fuel source, would require 92 cubic feet of air for the complete combustion of each pound of wood.

Because the degree of air tightness of the test containers could not be guaranteed, it was decided to use two 30 pound wood cribs per test. The total fuel source included 2 gallons of naphtha and occupied a volume of approximately 3 cubic feet.

The containers used for the interior fire scenario tests were removed from service because of their overall deteriorated condition. Minor defects, such as broken hinges, torn door gaskets, and dented frame rails, were noted on most of the containers. All containers were checked to insure that the defects would not produce any undesirable effects during the tests. No special repairs or sealing materials were employed to cause the containers to be air-tight beyond what is normal for in-service containers. In fact, these containers were probably less airtight than containers in normal service.

E.1.3.1 Thermocouple Locations

For Tests 1 through 3A, thermocouples were placed at various levels throughout each container. Two thermocouple "trees" were spaced approximately ten feet apart on the centerline of the container. Five thermocouples were mounted on each tree; one on the floor, one on the ceiling, and one every two feet in between. One thermocouple was placed in the center of the wood cribs, and one thermocouple was placed at the approximate midpoint of each side panel.

E1.3.2 Test Results

The ambient temperature and relative humidity for Tests 1 through 3A are listed in Table E-2.

Table E-2

Test No.	Ambient Temp (°C)	Relative Humidity (%)
1	35	50
2	38	54
3	36	65
3A	36	65

Table E-3 shows a summary of temperature and gas analysis data taken from the recorded curves. In Test 1, the fire became oxygen regulated in approximately 6 minutes, and subsequently, the temperature in the wood crib began decreasing with 2 minutes. In Test 2, the fire became oxygen regulated after approximately 9 minutes, however, the temperature at the wood crib did not show a corresponding drop for 72 minutes. It was noted, during set-up, that several of the floorboards were severely warped. It is theorized that fresh air entered through these openings in sufficient quantities to maintain glowing combustion in the wood crib, but insufficient to allow free burning. Tests 3 and 3A were identical except that the wood crib was placed at the curbside rear center of the container. See Section E1.2.5.1 for container details.

E1.3.3 Discussion

As noted in Section E1.3, the amount of available air in an empty sealed container is sufficient to sustain combustion and completely consume 14 pounds of wood. Test 1 involved a steel container with four natural vents. With this additional natural venting it was predicted that complete combustion of more than 14 pounds of wood could occur. A 13 pound wood crib weight loss was observed in the test. During the test, very little evidence was seen to indicate that the container's contents were burning. Minute traces of smoke were noted emanating from various points of the container. The only apparent effects of the wood crib fire on the container was a circular region, approximately 18 inches in diameter, on the ceiling of the container where the paint had burned off.

The FRP container (No. 5) used for Test 2 was not vented and still incurred a weight loss of 25 pounds. It is felt that the warped floorboards allowed sufficient quantities of air to leak into the container to permit glowing combustion of the

Table E-3. Summary of Data, Tests 1-3A

Test No.	Time (min)	Temperature (°C) Floor Level	2 Ft.	4 Ft.	6 Ft.	Ceiling	Wood-Crib	O_2 (%)	CO (%)	CO_2 (%)
1	2:00	36	35	37	41	50	32	21	0.2	0.5
(Stl)	4:00	41	39	41	49	58	290	21	0.2	0.5
	6:00	47	50	85	175	145	350	20	0.4	0.5
	7:00	–	–	–	180	170	400	–	–	1.5
	8:00	55	70	113	175	135	498	15	0.3	6.2
	10:00	–	–	–	–	–	–	16	0.4	5.0
	12:00	54	68	90	110	125	170	16	0.4	5.0
	16:00	53	63	75	90	92	145	16	0.4	5.0
	18:00	50	60	70	82	97	130	17	0.3	4.0
	20:00	42	48	58	65	87	105	18	0.2	1.0
2	3:00	35	35	35	35	35	75	21	0.2	1.2
(FRP)	5:00	–	–	–	–	125	–	21	0.2	1.2
	6:00	45	45	90	185	230	325	21	0.2	1.2
	7:00	–	–	180	–	200	350	20	0.4	1.2
	9:00	45	85	110	125	130	420	19	0.3	3.0
	17:00	–	–	–	–	–	275	14	0.6	4.2
	18:00	45	80	75	90	90	300	14	1.4	3.5
	21:00	–	–	–	–	–	290	12	–	–
	24:00	–	–	–	–	–	360	15	–	–
	27:00	45	80	75	90	90	375	14	1.6	3.5
	30:00	–	–	–	–	–	340	15	–	–
	33:00	–	–	–	–	–	375	12	–	–
	36:00	45	82	75	90	90	445	15	1.6	3.5
	39:00	–	–	–	–	–	475	15	–	–
	42:00	–	–	–	–	–	480	15	–	–
	45:00	45	82	78	90	90	510	15	2.0	4.0
	48:00	–	–	–	–	–	500	12	–	–
	51:00	–	–	–	–	–	515	15	–	–
	54:00	45	82	78	90	90	490	13	2.0	4.5
	57:00	–	–	–	–	–	470	14	–	–
	60:00	–	–	–	–	–	450	15	1.2	–
	63:00	45	81	75	90	90	425	–	1.2	3.8
	66:00	–	–	–	–	–	400	15.5	1.8	–
	69:00	–	–	–	–	–	375	–	1.7	–
	72:00	45	80	70	90	90	275	16	1.6	3.2
3	2:00	24	35	37	38	55	42	21	0.2	0.4
	4:00	36	38	44	64	66	485	21	0.2	0.4
	5:00	–	–	–	–	–	505	21	–	1.0
	6:00	42	55	110	165	114	530	20	0.3	2.0
	7:00	–	–	–	–	–	580	17	–	5.0
	8:00	54	85	142	194	142	388	14	0.4	6.0
	9:00	–	–	115	–	124	172	15	0.7	7.0
	12:00	52	72	145	102	102	132	17	0.35	4.0
A	1:00	–	–	–	–	–	40	–	–	–
(A1)	2:00	36	36	38	40	52	420	21	0.2	1.0
	3:00	–	–	–	–	–	480	21	0.2	1.0
	4:00	48	54	84	140	220	540	21	0.2	1.0
	5:00	–	–	144	–	304	560	16	0.3	4.0
	6:00	66	98	158	220	330	600	12	0.4	6.0
	7:00	–	102	156	–	304	500	13	0.4	8.2
	8:00	68	98	148	172	230	325	12	0.8	8.7
	10:00	64	90	118	118	165	200	15	0.5	5.5
	12:00	60	80	102	102	120	175	16	0.5	4.5

wood crib. Additionally, it was noted that the FRP panel joints were sealed with a material which apparently decomposed from the heat of the fire and leaked from the container. The liquid-decomposition-product that leaked from the joints may have allowed additional air leakage into the container.

There were two notable effects of the test fire on the FRP container. As in Test 1, a slight darkening of the ceiling occurred directly above the wood crib, and the heat of the fire caused a certain amount of styrene boilout on the roof panel during the test.

Test 3, involving aluminum container No. 6, produced results similar to Test 1. Prior to this test, it was predicted that the heat flux from the wood crib test fire would be sufficient to cause the aluminum roof panel to melt. The roof panel deformed inward approximately 1½ inches directly over the wood crib but did not melt. Apparently, the aluminum's thermal conductivity helped to dissipate the heat into the atmosphere. Table E-4 is a comparison of roof panel temperatures, of each type container, measured at the peak of combustion.

Table E-4. Roof Panel Temperature Comparisons

	FWD Thermocouple TREE (°C)	AFT Thermocouple TREE (°C)
Aluminum	140	140
FRP	240	240
Steel	170	180

Test 3A was conducted to determine if moving the wood crib to a corner of the container would produce any observable effects not produced in Test 3. The relatively undamaged container used for Test 3 was again used for this test. The wood crib was situated 5 inches from the curb-side-rear-corner of the container. The only additional effects noted were charring of the plywood liner, and the melting of the overhead door gasket.

In report number 177M, dated April 1973, the Netherlands Ship Research Center reported on a similar test series conducted in the Netherlands. In those tests, almost identical results were obtained for a ventilated steel container and an FRP container. In the Netherlands Ship Research Center tests, weight losses of approximately 12 pounds were recorded in all cases.

A second report, number 202M, dated November 1974, discussed the results of tests conducted on aluminum containers. As a result of these tests, the Netherlands Ship Research Center concluded that "a fire inside a container (regardless of the cause) will not inflict much damage on the container in question, and certainly not on adjacent containers."

E1.4 Exterior Fire Scenarios

In Section E1.1, it was stated that the two credible fire scenarios developed for intermodal container transport included exposure from an exterior fire source, either a minor flammable liquid leak or other source such as a fire in the container vessel's superstructure.

Tests 4 and 5 were designed to simulate the exposure of the top level of a container stack in such a manner that the underside of the containers would have no effect on the results. This was accomplished by flooding the test fuel pan with water to raise the fuel level above the container floors. The time of burning was regulated by the amount of fuel floated on the water's surface. Previous tests have shown that JP-5 fuel will burn off at a rate of approximately 1/10 inch per minute.

Since there was little damage incurred during the internal fire tests, the original three containers used for Tests 1 through 3A were also used for Tests 4 and 5.

Test 4 was planned as a 1-minute exposure. Unfortunately, immediately after ignition, a wind shift caused the fuel to move to the rear of the test pan. Consequently, the container doors experienced a 10-minute heat exposure. Although not part of the test plan, this provided an excellent opportunity to evaluate the container doors and locking systems.

E1.4.1 Test Results

The ambient temperature and relative humidity for Tests 4 and 5 are shown in Table E-5. Post-test examination revealed that all doors were still operable; however, a little extra effort was required to secure the locking rod cams in their keepers because the locking rods had bowed outward. The steel container (No. 4) had an S.I.C. type locking system, the FRP container (No. 5) had an Eberhard type locking system, and the aluminum container (No. 6) had a Fruehauf type locking system (similar to a Miner locking system). No failure of hinges occurred and all door gaskets were charred, but not totally destroyed. The aluminum container's Plymetal outer door panel (of aluminum sheet) had melted off, and the plywood core was charred.

The interiors of the containers were examined after Test 4 to evaluate the effects of the test fire. Other than discoloration from soot and charred door gaskets, no damage to the containers was noted. Since this test served to prevent the controlled escalation of fire exposure periods, and because the containers were still fairly intact, it was decided to fuel the test pan sufficiently to permit Test 5 to burn until complete destruction of the containers occurred. For this purpose, approximately 1¼ inches of JP-5 fuel was floated on the water to allow a 20-minute exposure.

Tables E-6 and E-7 show a summary of temperature data taken from the temperature curves recorded for Tests 4, 5 and 6, respectively. The temperatures were taken at different heights. See Section E1.2.5.1 for container details.

Table E-5

Test No.	Ambient Temp (°C)	Relative Humidity (%)
4	39	50
5	36	62

E1.4.2 Multi-Level External Exposure

Test 6 was intended to simulate a full-scale exposure of a container stack, in which the wooden floors of the containers are exposed. Nine containers were stacked in a three-by-three array (see Figure E-1). The fuel level in the test pan was lowered to approximately 6 inches below the floor boards of the first row of containers. Sufficient fuel was added to allow an exposure of approximately 25 minutes. The arrangement of containers and their respective construction materials are shown in Section E1.2.5.1.

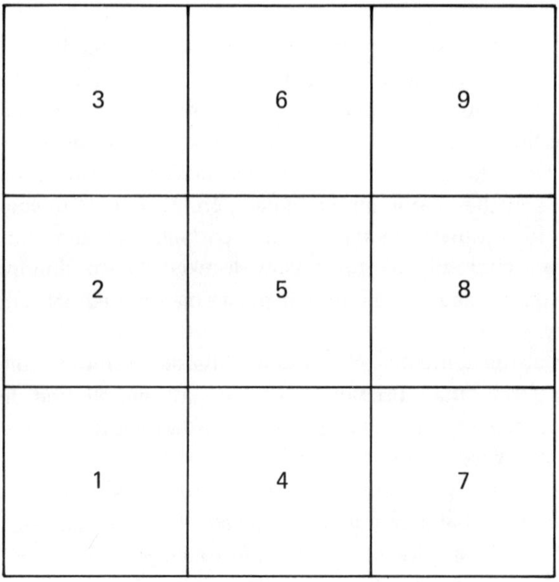

3	6	9
2	5	8
1	4	7

Figure E-1. Container stack 3 by 3 array.
NOTE: Refer to E1.2.5.1 for container types and numbers.

Table E-6. Summary of Data, Test 4 and 5

Test No.	Time (min)	Temperature (°C)				
		Floor Level	2 Ft. Level	4 Ft. Level	6 Ft. Level	Ceiling
4	2:00	39	39	41	41	66
(A1)	4:00	40	41	42	43	115
	8:00	41	42	44	46	100
	10:00	40	43	45	47	90
	12:00	41	44	47	48	80
4	2:00	34	36	37	37	42
(FRP)	4:00	33	37	38	38	53
	5:30	–	–	–	–	248
	8:00	36	41	42	45	110
	10:00	39	42	44	46	80
	12:00	40	44	45	47	72
4	2:00	40	39	47	45	57
(1)	4:00	98	114	133	150	230
	5:00	–	–	–	–	338
	6:00	290	338	350	360	330
	8:00	250	285	295	284	280
	10:00	200	239	242	244	230
	12:00	185	182	193	194	197
5	2:00	40	–	40	45	40
(A1)	4:00	82	–	60	168	175
	5:00	733	–	1020	1008	900
	5:30	945	–	940	975	920
	6:00	925	–	855	935	900
	7:00	897	–	840	800	775
	8:00	860	–	689	680	680
	9:00	897	–	920	795	775
	9:30	978	–	790	755	660
	10:00	898	–	710	765	690
	11:00	898	–	590	670	630
	12:00	730	–	525	510	460
	14:00	480	–	260	250	175
	16:00	460	–	190	240	160
	20:00	200	–	105	150	135
5	2:00	30	–	34	32	24
(FRP)	4:00	32	–	37	49	170
	6:00	38	–	54	78	140
	8:00	57	–	108	125	194
	10:00	84	–	170	230	342
	11:00	–	–	198	225	–
	12:00	100	–	184	210	220
	14:00	480	–	260	250	175
	16:00	460	–	190	240	160
	20:00	200	–	105	150	135

(Continued)

197

Table E-6. Summary of Data, Test 4 and 5 (Continued)

Test No.	Time (min)	Temperature (°C)				
		Floor Level	2 Ft. Level	4 Ft. Level	6 Ft. Level	Ceiling
5	2:00	30	–	34	32	24
(FRP)	4:00	32	–	37	49	170
	6:00	38	–	54	78	140
	8:00	57	–	108	125	194
	10:00	84	–	170	230	342
	11:00	–	–	198	225	–
	12:00	100	–	184	210	220
	14:00	114	–	170	195	179
	16:00	117	–	180	200	180
	18:00	132	–	195	224	194
	20:00	140	–	212	234	235
5	2:00	40	44	46	50	50
(Stl)	4:00	246	261	364	380	382
	6:00	410	470	475	490	460
	8:00	430	480	495	530	515
	10:00	470	515	515	555	550
	12:00	488	518	520	565	558
	14:00	490	520	522	560	545
	16:00	488	515	520	554	515
	18:00	435	470	480	530	490
	20:00	405	420	445	455	390

E1.4.3 Discussion

Above temperature of 450°F (232°C), is the approximate temperature above which aluminum loses its structural integrity. It is also the approximate kindling point for many types of Class A materials. Therefore, it is at this temperature that the spread of fire can be initiated, either by structural failure of container components, or by radiant or conducted heat energy. In Test 5, it was found that 450°F was reached in approximately 4 minutes in both the steel and aluminum containers, while the interior temperature of the FRP container did not reach this temperature for 9 minutes. The structural elements of the aluminum container began to melt after 4 minutes of exposure. However, the aluminum upper rail of the FRP container did not melt. At the end of the test, it was noted that the upper rail had begun to deform where it was connected to the steel end frame. It is felt that the insulating properties of the FRP side panels helped prevent a rapid transfer of heat to these components.

In Tests 4 and 5, the containers were free standing on the hatch cover. In Test 6, the containers were stacked and lashed together. This imposed a certain load upon

Table E-7. Summary of Data, Test 6

Container No.	Time	Temperature (°C)	Temperature (°C)
1	5:00	35	40
	10:00	210	157
	14:00	—	942
	15:00	635	628
	16:00	846	678
	20:00	110	40
	25:00	70	50
	30:00	65	55
	35:00	60	60
2	5:00	40	
	10:00	230	
	15:00	470	
	20:00	305	
	25:00	300	
	30:00	295	
	35:00	290	
3	5:00	40	
	10:00	145	
	15:00	145	
	20:00	745	
	25:00	820	
	30:00	745	
	35:00	625	
4	5:00	40	40
	10:00	185	596
	14:00	—	894
	15:00	550	670
	20:00	655	640
	25:00	650	640
	30:00	625	790
	35:00	550	894
	38:00	—	280
5	5:00	60	
	10:00	120	
	15:00	290	
	16:00	650	
	17:00	1080	
	18:00	1605	
	19:00	350	
	25:00	160	
	30:00	120	
	35:00	120	
6	15:00	47	
	10:00	47	
	15:00	320	
	16:00	600	
	17:00	740	
	18:00	780	
	19:00	920	
	20:00	990	
	25:00	960	
	30:00	910	
	35:00	855	

(Continued)

Table E-7. Summary of Data, Test 6 (Continued)

Container No.	Time	Temperature (°C)	Temperature (°C)
7	5:00	35	35
	10:00	195	130
	11:00	445	172
	12:00	520	245
	13:00	870	485
	14:00	230	216
	15:00	190	130
	16:00	580	216
	17:00	340	340
	18:00	720	216
	19:00	635	340
	20:00	770	780
	25:00	435	375
	30:00	290	420
	33:00	600	425
	35:00	290	430
8	5:00	45	
	10:00	655	
	11:00	800	
	12:00	650	
	13:00	810	
	14:00	966	
	15:00	695	
	16:00	805	
	17:00	850	
	18:00	910	
	19:00	800	
	20:00	650	
	25:00	695	
	30:00	655	
	35:00	610	
9	5:00	45	
	10:00	510	
	11:00	580	
	12:00	580	
	13:00	590	
	14:00	640	
	15:00	590	
	16:00	640	
	17:00	685	
	18:00	640	
	19:00	882	
	20:00	780	
	25:00	588	
	30:00	490	
	35:00	390	

(Concluded)

the bottom row of containers not experienced in previous tests. Additionally, the undersides of the container floorboards were exposed in Test 6. These two factors could account for the different results between Tests 5 and 6. The bottom row of containers in the stacked configuration of Test 6 did not reach 450°F for nearly 9 minutes. However, when this temperature was reached, containers 1 and 7 collapsed. The FRP containers had aluminum frames, while the remainder of the containers had steel frames, except container 5 which had an aluminum front end-frame. Throughout Test 6, the middle column of containers remained in place, with containers 3 and 9 also being held in place by the bridge fittings. The eventual failure of the container's aluminum front end caused the total collapse of the container stack.

E1.5 Conclusions

1. *Steel containers do not provide the utlimate degree of fire resistance.* The failure of steel side or end panels did not occur in any of the tests. The spread of fire through steel containers strictly as a result of direct flame impingement is therefore highly improbable. As a result of these tests, it can be shown that the transfer of a fixed amount of heat to the interior of a container from an external heat source will occur in approximately equal time periods for steel and plywood lined aluminum containers. A JP-5 pool fire source of approximately 30,000 Btu/ft.^{2}-hr. could cause the potential ignition of Class A materials inside a sealed steel or aluminum container in approximately 5 minutes. In Test 2, an FRP container delayed the transfer of this same amount of heat-flux for nearly 9 minutes. FRP containers can therefore provide optimum cargo protection for a short-term (9 minutes or less) external exposure.

2. *Extruded aluminum alloy frames do not provide an equivalent amount of integrity as high tensile strength steel frames.* In Test 6, the eventual collapse of the container stack was caused by the failure of aluminum frame components. Containers 1 and 7, which failed initially, utilized total aluminum frame hardware.

3. *The interior fire, in a sealed undamaged container, became oxygen regulated before any of the container panels are breached.* This concurs with results previously reported by the Netherlands Ship Research Center.

4. *The wooden flooring used in container construction does not add to the rapid spread of fire through container stack.* As discussed above, an external fire source will cause the transmission of heat through any of the three types of container panels in approximately 9 minutes, in an amount sufficient enough to ignite Class A materials. Laboratory tests using double thicknesses of nominal 1-inch tongue and groove flooring showed fire resistance ratings of from 12 to 18 minutes. In these laboratory tests (ASTM E-119), temperatures of 250°F (132°C) on the unexposed side of the floor was designated as a failure point. It can be rationally assumed then that the transfer of heat through 1-1/8- or 1-1/4-inch wooded flooring (in good repair) will require from 6 to 9 minutes to reach a point where the unexposed surface temperature approaches 450°F. This is essentially the same time as that required for the same amount of heat to be transferred through the container roof or side panels of all three container types.

5. *The stacking and lashing fittings currently used provide an adequate amount of structural stability under fire conditions.* In Test 6, the bridge fittings maintained the top row of containers in position even though the bottom end containers had collapsed.

E1.6 Summary

The findings indicate several important aspects of container shipping. An incident resulting in the ignition of cargo within a properly sealed non-damaged container, under most circumstances, will not endanger adjacent containers. However, a container stack exposed to an exterior fire source for more than approximately 5 minutes will most likely ignite and fail structurally causing flame spread to adjacent container stacks. External fire exposures produced nearly identical results in all types of containers. For this reason it is felt that changes in container construction are not necessary. Since containers do not act either to intensify a deck cargo fire or impede the spread of flame, a fire in an on-deck container stow will spread unless controlled by the installed fire protection system. It is essential that fixed firefighting systems be capable of rapid activation and application.

If not initially controlled, an on-deck container fire will progress until it exceeds the design application rate of the installed firefighting system, and at such time, the entire on-deck container load will be lost.

CHAPTER 8

SPECIAL SURFACE VESSELS
(FOILS, SURFACE EFFECT, ETC.)

8.1 Introduction

Vehicles in this category include hydrofoils, surface effect craft, and vehicles combining the attributes and operating modes of both types. This class of vehicle is characterized by light-weight construction in hulls, decking, and superstructure. Extensive use of aluminum is made in selected areas and, currently research is being conducted for application of advanced composite materials in foils, foil struts, underwater flaps, decking, and superstructure.

The special surface craft discussed in this chapter are many and varied in function, mission, size, and number of personnel aboard. Each possesses a variety of unique and peculiar features. However, they possess the common features of relatively light weight and high performance compared with other vehicles of their size. Structural loadings are such that non-metallics can be efficiently utilized in construction, and the mode of operation can vary from hydrodynamic hull support, air pressure support, or aerodynamic support.

The diversity of operation and construction represented by this class of vehicle is exemplified by the Patrol Craft Hydrofoil (PCH) and the Power Augmented Ram-Wing in Ground Effect (PAR-WIG) vehicles.

The PCH craft operates in the foil-borne mode between 40 and 50 knots. The superstructure, strut trunk housing, and sonar mast fairing are constructed of glass fiber reinforced plastic (GFRP). The hull is fabricated of a variety of aluminum alloys with the foils, struts, and flaps fabricated from HY-80 (High Yield) structural steel.

The PAR-WIG vehicle concept bears a close kinship to aircraft in mode of operation and construction. Operational requirements of this vehicle range from wing in ground effect operation to full flight at altitude to "sea setting." Materials utilized for structural components include aluminum and titanium alloys and advanced composite systems such as graphite/epoxy.

Hovercraft ride on an air cushion generated by the vehicle power system. Directional control can be effected by jet devices or propellers. This propulsive mode requires minimum weight in all system components to limit power requirements and vehicle maneuvering inertia. To obtain the minimum weight necessary for efficient operation, light-weight materials, both structural and non-structural, must be employed. The functional strength, stiffness, and environmental characteristics of polymer materials makes them prime candidates for such applications.

8.2 System Design and Operation

8.2.1 Introduction

Most special surface craft are completely self-contained. They provide all the mechanical resources and personnel support systems needed to safely and comfortably transport people or cargo over prescribed distances with reasonable economy and largely independent of other logistic and assistance sources.

In view of the nature and function of these vessels as well as the fire ignition sources and fire load levels present, fire hazards and control capabilities must be defined. Thus, this chapter defines the principal fire hazard modes for survivable incidents and discussed pre-fire initiation considerations, including determination of fire potential and current prevention and control systems.

It may be useful to consider here some characteristics of sea vessels that differentiate them from land structures, and make the use of noncombustible materials particularly important for special surface vessels.

1. The craft are self-contained. They must produce their own routine services as well as their own emergency services (water, electricity, etc.). Any fire may jeopardize essential services, placing such a vessel at the mercy of the sea.

2. Surface vessels are often remote from outside assistance. They must supply their own firefighting capability in order to maintain vessel integrity.

3. Means of escape are generally upward, the same direction fire and smoke spread most rapidly, rather than downward as in the case of buildings.

4. Weight limitations preclude the use of heavy, multi-hour fire barriers.

5. For safety and functional reasons, some vessels are divided into a number of discrete, moderate-sized compartments, making containment a practicable fire safety technique.

6. A salt and moisture laden atmosphere causes special hazards relating to electrical short-circuit and bi-metallic electrolysis.

These features make it essential to build a vessel with minimum fire load in order to limit the extent of fires in accommodation spaces. Recognizing that the contents of cabins (passenger and crew belongings, some furnishings, etc.) will be largely combustible, it is necessary to severely limit the amount of combustible material used in the construction of the vessel itself. Neither the presence nor absence of a material which is only noncombustible will affect the intensity or duration of a given fire (McDaniel, 1972).

8.2.2 Fire Hazard Modes for Survivable Fire Situations

Three types of survivable fire situations are considered: dockside/alongside fire, the en route fire, and the collision/loss of flight altitude fire. Dockside/alongside fires usually involve vessels at dockside or buoy with no passengers and few crewmen and small loss of life. History indicates that en route fires are most dangerous although fires following a collision or loss of flight altitude in surface effect (SES) or foil vessels could also be a serious problem.

A survivable fire accident is defined as an occurrence in which non-fire related injuries received by passengers and crew would allow survival of all or most of those persons.

8.2.3 Pre-Fire Initiation Considerations

8.2.3.1 Determination of Fire Potential

Ignition of polymeric materials is an extremely complex process (see Chapter 4). It depends on the nature and characteristics of the ignition sources, the availability of adequate oxygen, and the physical and chemical properties of the polymer. Important properties of the polymer influencing its "ignitability" include thermal conductivity, density, thickness, specific heat, and activation energy. Once ignited, polymers may burn intensely, releasing large amounts of energy. This energy is transferred by radiation and convection to other combustible materials, causing ignition and propagation of the fire.

Some polymeric materials maintain their structural integrity as they burn, while others melt and sag. The latter may constitute a greater hazard when they are used in load-bearing applications rather than in decorative ones. Other characteristics (e.g., melting and dripping, smoke evolution, rate of heat release, and burning rates) are additional concerns that can be evaluated only in a finished product. The best evaluation for these materials is in their "use" configuration.

Time is a critical element in a fire; an increased effort should be made to learn how to acquire more time for a given ignition intensity. Extinguishment characteristics also need to be defined in sufficient depth. Flashover is an as yet insufficiently explored aspect of fire dynamics which involves area geometry, ventilation, and vapor accumulation from pyrolyzing organic materials (see Chapter 3) and should be designed to obviate it as an ignition source even in case of catastrophe. Parts that are electrically insulated from the hull structure can constitute a hazard unless connected to it through proper ground or shielding.

Fires may be initiated as a result of heavy current overload, fraying of insulation, or breaking of the wire. Insulation of the wire should be selected with regard to fire retardance and the products of combustion of the insulation. In addition, wiring should not be installed to permit contact with flammable fluid lines. In areas where wiring must be located in close proximity to fuel lines, the wiring support should be located well above the lines to prevent the loose end of a broken wire from contacting flammable fluid pipes. (It has been shown by tests that in such cases an arc may burn through the line and set fire to its contents before circuit protection has time to act).

Terminals must be covered to prevent accidental shorting or grounding, and all plugs and receptacles should be sealed when locked in the connecting position. The seal should be established before electrical contact is made and be maintained until after contact is broken.

Circuit protection must be provided for all circuits exposed to transient current input in excess of normal wire rating. This circuit protection may either be installed outside abnormal vapor zones or be explosion-proofed.

Main power cables (including generator cables) should be: (1) isolated from flammable fluid lines, (2) shrouded by electrically insulated flexible conduit or its equivalent, in addition to the normal cable insulation, and (3) designed to allow a reasonable degree of deformation and stretching without failure.

The salt laden high-moisture atmosphere in which the equipment operates exacerbates and will accelerate the deleterious effects of the hazards noted above. (Refer to item 6, Section 8.2.1).

8.2.3.1.1.2 Cigarettes, Matches, Lighters

Temperatures achieved by small heat sources (e.g., cigarettes, matches, lighters) are sufficient to ignite many materials, including some synthetic polymers, both in solid and liquid form. Whether or not these sources will ignite a given material depends on the material configuration and the atmospheric dynamics.

The following are some typical temperature measurements of small ignition sources:

Item	Condition	Temperature
Cigarettes, center	No draft	1,050°F (565°C)
Cigarettes, center	Draft	1,350°F (732°C)
Cigarettes, center	Insulated	1,150°F (621°C)
Cigarettes, surface	No draft	550°F (228°C)
Cigarettes, surface	Draft	800°F (427°C)
Paper Match	No draft	1,508°F (820°C)
Wood Match	No draft	1,346°F (730°C)
Cigarette Lighters	No draft	1,200–1,500°F (649–816°C)

Testing a typical cross section of upholstery materials using a cigarette and other small ignition sources revealed that combinations of cotton and rayon materials were ignited by these sources, while certain 100-percent synthetic fibers were not (Loftus, 1956) (National Fire Protection Association, 1956). Synthetic materials burn when blends of cotton and rayon fibers are combined with synthetic fibers or where cotton is used as a cushioning material, since the easily ignited material serves to ignite the more difficultly ignitible materials. It should also be pointed out that ignition by cigarette generally begins with a smoldering phase so that considerable time may elapse between ignition, flame growth, and flame spread.

Many design techniques have been developed to reduce ignition by cigarette. These include incorporating a heat sink material (e.g., an aluminized scrim material beneath a cotton-polyester material will inhibit cigarette ignition), and using tight weave designs with high-density fabrics.

8.2.3.1.2 Fireload

One classic approach to understanding the potential severity of fires in a given space has been to measure "fire load" or potential heat release under fire conditions.

Every potential fire in a vehicle — one that could occur at any time or place, whether it is in operation or stored, or with full load or empty — must be considered one in which all combustibles will be consumed. The question is what will this total heat load do if it is released in a relatively short time?

Heats of combustions of various materials that can find application on this class of vessel are listed in Table 8-1.

Table 8-1. Material Fire Load Data

Heats of Combustion for Typical Materials Found in Special Surface Vessels

Rayon	6,700 (Btu/lb.)
Tobacco	6,800 (Btu/lb.)
Cotton	7,100 (Btu/lb.)
Acetate	7,700 (Btu/lb.)
Polyvinyl Chloride	7,700 (Btu/lb.)
Tri Acetate	7,800 (Btu/lb.)
Wood	8,800 (Btu/lb.)
Wool	9,000 (Btu/lb.)
Polyester	9,300 (Btu/lb.)
Modacrylic	11,000 (Btu/lb.)
Unsaturated Polyester	13,000 (Btu/lb.)
Nylon 6	13,000 (Btu/lb.)
Spandex	14,000 (Btu/lb.)
Foam Rubber	15,000 (Btu/lb.)
*Bituminous Coal	15,000 (Btu/lb.)
Urethane	16,000 (Btu/lb.)
Polystyrene	18,000 (Btu/lb.)
*No. 6 Fuel Oil	18,000 (Btu/lb.)
Butadiene/Styrene Copolymer	20,000 (Btu/lb.)
*for comparison purpose only.	

8.2.3.1.3 Personnel Danger

The crew may be involved in many types of fire, either by the necessity to continue the vehicle's operation or to fight the fire itself.

The use of all fire-retarded cellulosic types of fiber in crew clothing would reduce the probability of ignition but would not ensure a protective garment; the use of Kynol®, Nomex or fiberglass would confer additional fire protection. The use of hooded capes, thermal gloves, and a self-contained breathing apparatus would add to safety.

8.2.3.2 Current Fire Prevention Control Systems

8.2.3.2.1 Prevention

Fire prevention starts with vehicle design and follows through by considering all potential operational accidents. Ignition sources are (a) eliminated or controlled to the extent possible and (b) separated from possible fuels. Fire detection systems that provide for quick operation of fire suppression systems before a fire builds up are installed in engine room and ammunition magazines and in unoccupied areas of the vehicles.

There are no effective controls on the materials or ignition sources carried on board by passengers or crew.

The crew and passengers are expected to detect a fire, and the crew is expected to put it out using available extinguishing systems before it is too large to control.

Some surface vehicles are of a size such that barriers are used to separate cargo, crew, and passenger compartments. Such separations also help to contain fires; however, consideration by designers of the fire barrier concept beyond this has been minimal. Servicing of fuel, hydraulic, and oxygen systems presents such severe fire hazards that special conditions, facilities, and procedures are required to prevent fire.

In essence, fire prevention control and extinguishment must be treated as a system problem since all parts of the fire problem are related (e.g., tradeoffs in safety and economics must be made in using materials that are difficult to ignite but that when ignited, produce fire, smoke, or toxic gas conditions).

8.2.3.3 Design Considerations

8.2.3.3.1 Introduction

The purpose of this section is to discuss the design of polymeric material components, insofar as the material selection is relevant to fire safety. Material selection is considered as it pertains to hardware parameters controlled by the designer. These parameters include part function, geometry, location, and the influence on materials of service temperatures.

8.2.3.3.2 Component Considerations for Material Selection

8.2.3.3.2.1 Part Function

All hardware items have a primary purpose or function. The means to satisfy this function is through a requirement specification, i.e., a design criteria document. Once the design criteria have been established, the design process is initiated. This usually involves a series of tradeoff studies, frequently resulting in compromises among requirements. If function cannot be provided or the required criteria met, consideration of other desirable features becomes academic. This premise establishes the sequence of design flow for any hardware element regardless of second- or third-order requirements of desirable features that provide an additional basis for tradeoff study selection of materials for a given application. Flow sequence for candidate component design relative to function and fire characteristics is illustrated in Figure 8-1.

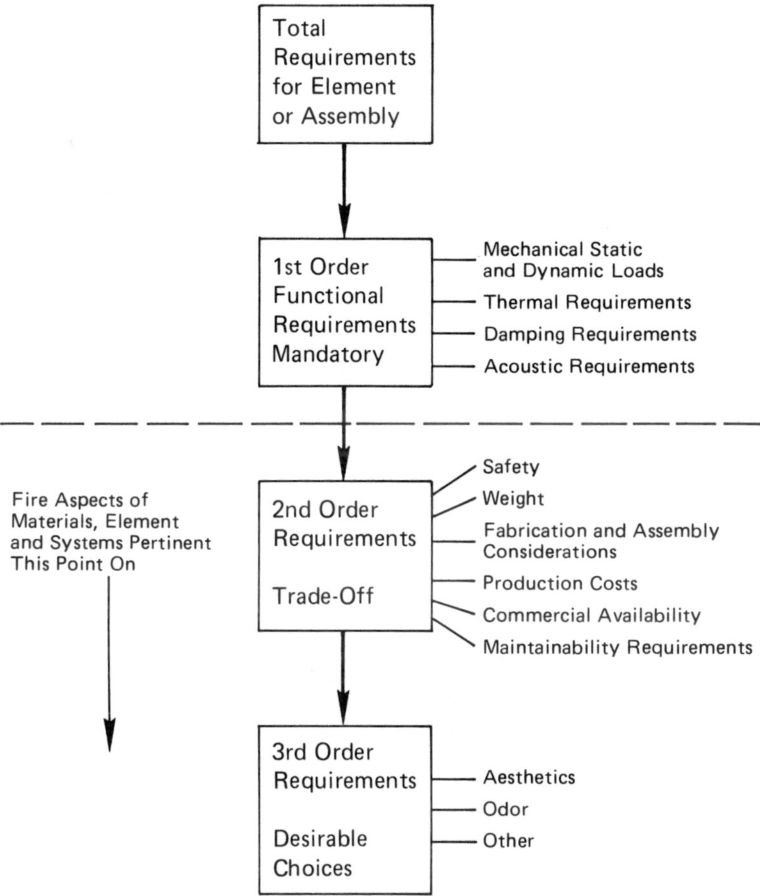

Figure 8-1. Flow diagram for candidate componentry (interior nonstructural bulk-heads, deck covering, galley hardware, paneling, habitability items).

8.2.3.3.2.2 Geometry

The geometry of interior hardware (interior non-structural bulkheads, seats, deck covering, galley hardware, paneling transparencies and ports, comfort items, and lavatory fixtures) is very important in terms of fire flashover, and particularly in terms of retention of heated gases and the focusing of explosively ignited gases. Flame propagation rates vary according to the path of the flame front (i.e., verti-cally, horizontally, or at some inermediate angle). These geometric considerations are difficult to control meaingfully, but should be given attention if design function can be accommodated.

8.2.3.3.2.3 Part Location

The location of a part can constitute an essential element in material selection by virtue of the response of the available polymers to the elements of a given fire scenario, i.e., a material's propensity to drip, emit smoke and toxic fumes, self-extinguish, and respond to heat flux. Other significant properties of a polymer, as they relate to its combustion performance, can also be extremely important from the standpoint of fire propagation or retardation in a given scenario. Materials that melt and drip are extremely hazardous to components if located in a position that permits them to ignite other materials.

Location of a part relative to adjacent high temperatures, high rate of energy input, environment (particularly partial pressure of oxygen), and heat loss are important parameters for its ignition and sustained burning. Ignition testing of fibrous materials with a bunsen burner has demonstrated that some materials which self-extinguish at room temperature will be completely consumed if the ambient temperature is elevated to 250°F (120°C). Data for a group of materials tested using the Federal Standard Test Method (FSTM) method are presented in Table 8-2.

Table 8-2. Bunsen Burner Test. Elevated Temperature Versus Room Temperature

Material	Room Temperature Air[a]		250°F Air[a]	
	Flame Time, s.	Burn Length, in.	Flame Time, s.	Burn Length, in.
Nomex® fabric, undyed	0	2.5	0	3.3
Wool fabric, FR treated	0	2.5	9	10.0[b]
Nomex® carpet, polyester back, FR latex coating	17	0.6	3.8 min	8.8[b]
Wool carpet, polyester back, FR latex coating	8	1.8	48	8.8[b]
Wool carpet	15	3.6	2.4 min	10.0[b]
Dynel carpet	5	3.3	5	10.0[b]

[a]Temperature of air in burn chamber.
[b]Specimen was fully consumed.

8.2.3.3.2.4 Detail vs Assembly

Proper evaluation of the combustion quality of a material in a given application requires consideration of the material or combination of materials in an assembly and the material in the adjacent assembly.

Surface texture, color, shape, weave, gauge, density, etc., affect the ignition, burn, and smoke characteristics of otherwise chemically identical materials when tested by the same methods. Therefore, detail parts cannot be averaged to obtain their combined characteristics but must be tested as the complete end hardware assembly. This is exemplified and the data presented in Figure 8-2. With this combination of materials, in vertical burn tests, the burn length of the complete panel assembly averages 1.1 inch longer than that resulting in an identical test conducted on a phenolic resin/glass laminate.

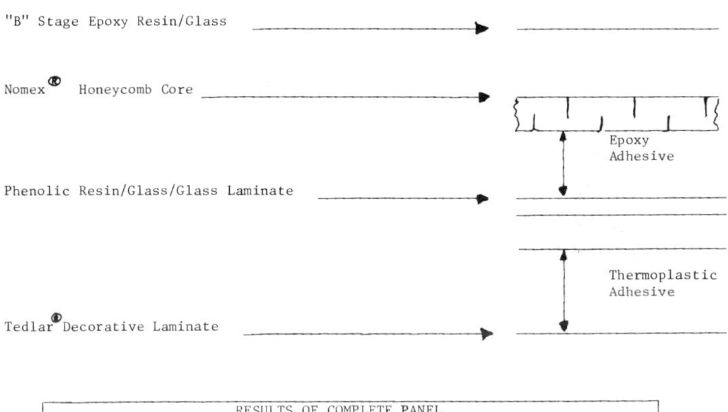

RESULTS OF COMPLETE PANEL ASSEMBLY		
Flame Time (Secs)	Burn Length (Inches)	Flame Time of Drippings (Secs)
0	4.70	No Drip
0	4.90	No Drip
2	4.50	No Drip
AVG. 1	4.70	No Drip

RESULTS OF PHENOLIC RESIN/GLASS LAMINATE		
0	3.60	No Drip
0	3.60	No Drip
0	3.60	No Drip
AVG. 0	3.60	No Drip

Figure 8.2. Response to heat flux component detail versus assembly. Sidewall panel.

8.3 Materials

8.3.1 Introduction

Materials used in this special class of vehicles must be specified with full consideration for total system requirements including ignition sources, fire detection, and fire control.

Because operation of these special craft poses unique problems, polymeric materials used should have the highest fire safety performance consistent with pragmatic limitations of availability, cost, and other performance properties (e.g., mechanical durability, weight, aesthetic appeal, mar resistance, ease of cleaning, and fabrication).

Area designated by the Coast Guard and Navy as having special fire safety importance are:

1. Areas or compartments not normally manned.
2. Areas occupied by passengers and crew (i.e., work stations and living quarters).
3. Areas such as the hull where polymeric materials (e.g., linings, wire insulation) are important components.

The first two are interior areas, and polymeric materials are utilized in several applications including:

1. Thermal and acoustical insulation, bulkheads and overheads, survival gear, and oxygen systems;
2. Seats, floor covering, mattresses, draperies, pillows, and blankets;
3. Passenger and crew carry-on materials.

In some instances, military vehicles are made of materials of the same type and form as found in commercial craft, the principal difference being the elimination of some materials required for interior decoration. Military vehicles are more limited in the use of polymeric materials, generally retaining only those essential to the system. The habitability materials for carpeting, curtains, mattresses and upholstery fabrics burn, smolder, and give off toxic fumes. Fuels, explosives, metals and plastic materials all compound the danger of shipboard fire.

In conventional Navy ships and advanced high-performance ships, some materials and equipment contain substances such as aviation and rocket fuels. These explosive materials, common to the working environment within many naval ships, pose serious problems. Lightweight structural materials such as aluminum, replacing the traditional heavy steels, bronze and copper, are used for hulls, helicopter pads, and hangars. Aluminum has its drawbacks, however, as it begins to lose strength at 250°F (121°C) and melts at 1184°F (640°C), a temperature much lower than produced by shipboard fires.

Advanced composite materials are replacing aluminum in some applications and may possibly replace steel in selected applications. Recent experimental results have indicated that some composites possess an apparently higher level of resistance to burn-through than certain aluminum structures when subjected to direct flame fronts at 1093°C (2000°F). The subsequent pyrolysis and burning of the polymeric resin matrix of the composite will constitute a severe hazard, however.

Current test methods specified by government regulatory agencies are based on technical developments of the scientific and industrial communities. They have been used successfully to develop substantially safer vessels by permitting replacement of highly flammable materials with improved materials. However, these test methods usually measure a limited number of individual characteristics and do not provide an integrated assessment of all fire safety parameters. The standards were developed from tests reflecting what was the state-of-the-art at the time, and usually do not take into account aging phenomena or design considerations. Furthermore, correlation of test results with actual fire situations is difficult.

Many polymeric materials are used in blends with other polymers or as composite structures. The fire safety performance of these structures is complex and difficult to predict, since the evaluation of individual components does not provide complete information about the performance of the composite or multi-component system during exposure to fire.

Since most special-purpose, high-performance craft are experimental or developmental, many different materials and composites are specified or allowed. These materials may be called out in accordance with conventional ship construction practice, aircraft construction practice, or in accordance with the requirements of a materials development program. Since the materials specified for surface effect, foil, PARWIG and other special purpose craft vary widely, even for the same functional part, it does not appear useful to attempt to describe and evaluate here all of the many polymers and composites in use. Therefore, this chapter is limited to a discussion of the design principles and criteria for materials selection from a fire safety standpoint. Materials used in aircraft construction are discussed extensively in Volume 6 of this series; ship construction materials are discussed in Chapters 4 and 7 of this volume.

Materials carried on board by or for passengers and crew cover a wide range of liquids, solids, aerosols, and other compositions of varying degrees of potential fire hazard. Many of these materials are substantially more flammable than the construction and furnishing materials used by the vessel manufacturers and operators. They constitute a significant factor in the overall fire safety of the vessel — one that is not susceptible to much improvement except by containment and, to some degree, by regulation. Significantly, these materials kindle less flammable materials.

To date, the concern for the fire safety aspect of material selection has focused on resistance to ignition or prompt self-extinguishment; however, the potential smoke generation and toxic gas evolution characteristics of the selected material now are receiving attention. Evidence is accumulating suggesting that many presently used self-extinguishing formulations of polymeric materials cause increased smoke and toxic gas release in a conflagration.

8.3.2 Evaluation Criteria and Methodology

Evaluation and selection of polymeric materials is a complex and difficult task. It is not now possible to precisely define fire safety without reference to specific parameters or conditions of testing. The fire safety aspects of a polymer depend on many factors including actual condition of use. This is particularly true of the geometry and orientation of use, proximity of other materials, environmental conditions, and source and site of ignition, as well as the intrinsic properties of the polymer such as composition, thermal stability, and heat transfer characteristics. As noted previously, the effects of decomposition products (smoke and toxic gases) must also be considered. Results can be interpreted only with reference to the test procedures employed and with full awareness of the prevailing limitations in test methodology.

A further distinction must be made with regard to the intent of various flammability test methods. In the hierarchy of procedures used for materials selection, distinction should be made among tests and the results should be meaningfully and carefully qualified. Investigation of evaluation methods has led to identification of performance criteria that are related to fire safety (see Chapter 3 and Chapter 4).

In summary, the methodology available and the evaluation criteria selected for assessing the fire safety of specific materials to be used in the special vessels are only qualitative and tentative; they are presented, with full awareness of their limitations, as a stepping stone for future progress. A continuing reevaluation of these materials in view of new methodology and knowledge as they become available is mandatory.

8.3.3 Fire Retardation of Polymers

The development of polymers with inherent thermal stability has been emphasized in attempts to reduce flammability of aircraft materials. It can be expected that special vessels will be enhanced by direct technology transfer. See Chapter 4 for a more detailed treatment of fire retardation.

8.3.4 Application of Polymers

Polymers are used extensively throughout this special class of surface vessel for hulls mechanical items, paint, thermal and acoustic requirements, and furnishings. The polymers used are both natural and synthetic.

The many contemporary uses of polymers can be put into several broad categories: mechanical items (gaskets, packings, hoses, seals, flexible connections), electrical insulation, thermal insulation, noise suppression, protective coverings (paints, deck coverings), furnishings (bedding, chairs), protective clothing (gas masks, fire fighting suits), life saving equipment (inflatable life boats, life vests), and packaging and packing material.

Material selection factors for a given application are primarily function, availability, and life-cycle cost. Additional factors such as weight saving, user acceptance, and environmental resistance also influence the choice (refer to Navy Habitability Guidance List of Acceptable Materials, Rev. D, 1977, MIL-STD-1623B).

8.3.4.1 Special Problems

Many of the craft in this special class of vehicle differ considerably from the common concept of water borne vessel. The construction materials, mode of operation, vehicle weights, operating speeds, and internal configuration are more closely allied to those of aircraft than ships or other water craft. These differences add complexity and a level of uncertainty as to vulnerability to fire as well as the ability of the craft's systems to fight fire. Personnel escape modes vary among vehicle types in this class but all differ from classic water borne vessel emergency egress methods. Because of the hybrid nature of this vehicle class, it could be possible to overlook some problem areas heretofore not uncovered or experienced by existing craft.

8.4 Tests

8.4.1 Introduction

As noted, polymers are utilized as construction materials and functional hardware materials for this class of vehicle. Vessels in this class have, by their function, an extremely high premium placed on weight reduction. This results in a drive toward extensive use of lightweight materials such as aluminum and fiber-reinforced plastics for decks, hulls, and superstructures, including interior bulkheads and partitions. Material applications, as described above, are subject to both combustion and burn-through requirements.

Both the Coast Guard and the Navy have vessels of this type under their cognizance, with attendant test methods, specifications, and standards. Test descriptions are contained in Chapter 5.

8.4.2 Navy Tests

Most fire tests and requirements are codified in Military Standard 1623, Fire Performance Requirements, and Approved Specifications for Interior Finish Materials and Furnishings (Naval Shipboard Use). These requirements are for minimum flame spread and smoke developed by standard tests such as ASTM E-84, ASTM E-162 and the NBS Smoke Chamber. Materials with a flame spread greater than 25 by either ASTM E-84 or E-162 are not utilized if possible. Smoke requirements are not specific, but the goal is to be not over 100 by the NBS method, D corrected (See Chapter 5).

Wire and cable are mostly procured to MIL C-915, which requires a simple bunsen burner test to determine fire propagation of a single wire or cable. This has proved to be an unsatisfactory test, as multi-wire cable runs will burn with intensity and propagate fire. There are tests available that appear more appropriate for this material application, including the IEEE 383 test and the heated coil test of ASTM D-229 which is a relatively severe test.

Glass fiber reinforced plastics applications are covered in the resin specification, MIL-R-21607, which essentially uses the hot coil test of ASTM D-229 (see Chapter 5).

8.4.3 Coast Guard Tests

The Coast Guard is responsible for development of fire protection requirements related to construction and operation of merchant vessels. In effect, the merchant vessel construction regulations are a "Marine Building and Exits Code." These regulations may be found in title 46, Code of Federal Regulations.

8.5 Smoke and Toxicity

8.5.1 Introduction

Major factors that influence survival of persons subjected to fire environment in confined spaces are:

1. Heat destruction of tissues due to thermal shock.
2. Toxicity from oxygen deficiency, exposure to carbon monoxide and other noxious gases, aerosols, and particulate materials.
3. Presence of smoke with consequent reduction of vision and visibility.
4. Fear or outright panic resulting in secondary mechanical trauma.

All of these factors may be involved in a fire, depending on the fire scenario and the individuals in the confined space. The nature, shape, and quantity of materials undergoing combustion or pyrolysis determine the number of factors involved in any fire and therefore the degree of hazard to human survival.

The types of vehicles discussed in this chapter are examples of a confined space in which fire represents a serious hazard. The material that undergoes combustion or pyrolysis (excluding, for the purpose of this report, engine fuel, lubricants, and hydraulic fluids) consists of natural and synthetic polymers. In general, with the exception of wool, cotton, paper, and other cellulosic materials, the most flammable materials on board these vessels are synthetic polymer materials.

The smoke and toxicity problem is described in more detail in Chapter 6.

8.6 Conclusions and Recommendations

Conclusion: High-performance vessels require minimum weight to achieve performance, and polymeric materials offer great promise in this area. Most polymeric materials used aboard special high-performance vessels are easily ignitable and produce heavy volumes of toxic gases and smoke. *Recommendation:* Known hazardous polymeric materials currently used in inhabited areas of dynamically supported vessels should be replaced with available improved materials.

Conclusion: Vessel crews may be involved in fires in which it is necessary to continue the operation of the vessel while fighting the fire. *Recommendation:* Vessel crews should be required to wear regular clothing fabricated from commercially available fire-resistant fibers (e.g., treated wool, treated cotton, treated polyesters, aramids, and phenolics) while on duty to provide increased protection against fire.

REFERENCES

Bogner, C., Materials Research For Improved Fire Protection, Unpublished, July 1977.
Loftus, J., "Unpublished Work," National Bureau of Standards, Washington, D.C., 1956.

McDaniel, D. E., "Noncombustibility: A Marine View," ASTM Reprint, Philadelphia 1972.

National Fire Protection Association, "Cigarette Fire Mechanism," Boston, Massachusetts: 1956.

Pratt, D., Presentation to NMAB Committee on Fire Safety Aspects of Polymeric Materials on November 30, 1976.

Sheehan, D. F., Presentation — NMAB Committee on Fire Safety Aspects of Polymers, November 1976.

Sheehan, D. F., "Progressive Fire Protection," The Society of Naval Architects and Marine Engineers, Williamsburg, Va., May 1972.

CHAPTER 9

BOATS AND CRAFT

9.1 Introduction

9.1.1 Scope and Applications

The focus of this chapter is the fire safety of polymeric materials on small boats not more than 65 feet long, propelled by sail or power.

Most boats in this category are used primarily for recreational purposes. Some may be used in commerce for transporting passengers, as cargo utility boats, or for commercial fishing. Others will be used for specialized functions such as police patrol, firefighting, customs, and hydrographic surveying. The U.S. Navy uses these craft as personnel carriers, utility boats, tenders, and river patrol boats.

The boats covered in this chapter and covered in other parts of this volume are self-contained units. They must provide their own routine utility services; they must also fight their own fires. Escape from fire is to the open water, generally with the benefit of no more than a life jacket.

The vast majority of the 9 million boats are small and built for the mass market. The cost constraints of severe competition demand that the materials and the fabrication be competitively priced. The boats are generally fabricated from materials based on unsaturated polyester resins. See 9.2 for a discussion of the economics of hull building with fire retardant resins. These constraints usually limit the use of heavy and costly fire safe construction.

Nevertheless, there may be specific applications in which the fire risk is high and the fire safety benefit to be gained so great that substitution of more expensive material is justified; or, in the case of an extreme risk, substitution could be mandated by standards and regulation.

9.1.2 Statistics on Life and Property Loss Due to Fire

The U.S. Coast Guard registry and casualty statistics (Boating Statistics, 1976) show that in 1975 there were almost 9 million pleasure and small commercial boats registered in the U.S. Of that number, 8,002 (Table 9-1) were involved in accidents that caused 1,466 fatalities and 2,136 other injuries. Twelve of the fatalities and 261 of the injuries were related to fire or explosion (Table 9-2). Fire appears to be a relatively small factor in personal injury and death aboard boats.

The same statistics show a different story of property damage as a result of boating accidents. In 1975, 5,339 boats reported $10.4 million in damages. Although the total loss is small, 380 boats accounted for $3.3 million, about a third of the total. The average loss per reported incident is about $5,000. Fire accounts for about 30 percent of the losses. The figures suggest that a large portion of the reported fire accidents resulted in total destruction of the boat.

Table 9-1. Vessels Involved in Accidents

Types of Casualty	Total					Fatal					Injury					Property Damage				
	1971	1972	1973	1974	1975	1971	1972	1973	1974	1975	1971	1972	1973	1974	1975	1971	1972	1973	1974	1975
Grounding	210	240	283	284	437	11	12	17	11	11	14	38	39	28	67	185	190	227	245	350
Capsizing	641	623	874	778	751	474	434	603	455	459	28	18	71	39	77	139	163	100	284	815
Flooding/Swamping	173	152	149	154	256	64	62	35	30	61	4	5	8	9	18	105	85	106	115	277
Sinking	177	203	335	351	289	48	44	66	61	49	7	8	15	23	11	130	151	254	277	238
Fire or explosion of fuel	350	309	372	366	473	15	15	14	10	10	78	53	101	58	140	257	241	257	298	323
Other fire or explosion	77	48	105	85	70	2	4	3	4		2	4	20	3	13	73	40	82	78	57
Collision with another vessel	2009	2261	2853	2845	3534	87	97	100	80	53	378	270	537	307	463	1544	1894	2216	2458	3018
Collision with fixed object	415	338	599	573	752	53	46	60	62	67	91	58	132	112	165	271	234	407	399	520
Striking floating object	164	173	190	155	292	15	31	18	15	13	20	11	29	21	40	129	131	143	119	239
Falls overboard	352	375	433	362	421	314	317	364	312	301	32	49	62	37	100	6	9	7	13	20
Falls within boat	10	3	36	39	48	1		5	10	4	8	3	30	24	36	1		1	5	8
Struck by boat or propeller	46	72	109	100	138	13	17	15	21	13	24	55	92	69	104	9		2	10	21
Other casualty; unknown	291	255	400	357	541	168	118	166	134	146	71	75	168	106	251	52	62	66	117	144
TOTAL	4915	5044	6738	6449	2002	1257	1197	1466	1205	1178	757	667	1304	826	1405	2901	3200	3968	4418	5339

Table 9-2. Results of Boating Accidents

Types of Casualty	Fatalities					Injuries					Amount of Damage (dollars)				
	1971	1972	1973	1974	1975	1971	1972	1973	1974	1975	1971	1972	1973	1974	1975
Grounding	15	15	20	14	13	20	53	70	37	104	1,040,000	867,000	1,752,200	817,600	1,261,400
Capsizing	659	574	796	602	609	74	80	137	79	155	692,000	350,000	429,100	692,900	357,500
Flooding/Swamping	82	81	48	38	86	8	13	11	14	32	323,000	355,000	264,300	406,700	451,800
Sinking	63	68	89	80	59	12	15	34	26	20	534,000	720,000	901,400	789,800	760,400
Fire or explosion of fuel	18	16	18	12	12	123	97	186	96	239	2,704,000	1,884,000	3,765,900	2,448,900	2,493,000
Other fire or explosion	2	4	4	5		4	5	27	7	22	1,767,000	315,000	869,100	647,700	758,900
Collision with another vessel	83	64	67	53	66	350	237	449	272	673	987,000	1,636,000	1,947,200	1,728,300	2,190,900
Collision with fixed object	61	55	65	73	79	120	107	238	165	297	568,000	410,000	1,021,400	1,054,800	1,062,100
Striking floating object	20	37	21	17	13	22	26	37	32	57	280,000	326,000	255,800	287,500	528,500
Falls overboard	336	337	390	330	317	38	58	85	43	100	17,000	9,000	25,200	19,200	30,000
Falls within boat	1		9	10	4	10	3	34	26	42			11,600	6,800	12,300
Struck by boat or propeller	13	16	15	21	14	24	57	97	73	110	9,000		2,900	6,000	14,900
Other casualty; unknown	229	170	212	191	194	92	78	194	123	277	101,000	235,000	131,500	276,200	240,400
TOTAL	1582	1437	1754	1446	1466	897	829	1599	993	2136	9,022,000	7,107,000	11,374,600	9,181,500	10,352,200

9.1.3 Statistics on Fire Losses versus Hull Material and Propulsion Systems

Examination of the Coast Guard statistics for 1975 shows the propensity toward fire accidents to be dependent on hull material of the boat and on the method of propulsion. Thus, Table 9-3 shows that vessels with wood hull are involved in fire or explosion incidents more than 2½ times as often as would be expected from their proportion in the boat population; fiberglass hulls are involved in fires one-third more often than their expected proportion; and aluminum hull boats are only one thirteenth of their expected proportion.

Table 9-3. Hull Material Vs. Fire Incidents Numbered Motorboards Under 65 Feet

Hull Material	(1) Percent of 6.9 million Numbered Boats	(2) Actual	(3) Expected*
		Vessels Involved in Fire/Explosion	
Wood	14%	93	35
Fiberglass	43	143	107
Aluminum	37	7	92
Steel	2	5	5
Other	4	0	10
Unknown	-	1	-
Totals	100%	249	249

*Expected fire/explosion incidents
= Column (1) X Total of Column (2) = Column (1) X 249

Table 9-4 shows that, whereas outboard propulsion boats are 86 percent of those registered, they are involved in only 18 percent of the reported fire/explosion incidents.

Table 9-4. Propulsion Systems of Boats Involved in Fire/Explosions 1975

Propulsion*	No. of Vessels	% of Vessels Involved
Outboard	45	18
Inboard-gasoline	108	43
Inboard-diesel	23	9
Inboard-Outboard	66	26
Other	4	2
Unknown	3	1
	249	100%

*1975 Classification of all numbered motorboats under 65 ft. was: Outboard 86%; Inboard 14%.

9.2 System Design: Structural Hull

Only in the very recent past has wood as a boat building material been successfully challenged by a variety of metals, ferrocement products, and recently, polymers, generally in the form of fiberglass reinforced polyester (FRP). Each of these materials has its own inherent advantages — as well as disadvantages.

The wide acceptance of current FRP boat construction is the result of the low cost of simplified techniques for the builder on one hand, and ease of maintenance for the owner on the other. Fiberglass boat builders no longer have need for large numbers of highly skilled carpenters and joiners so much in demand in the construction and finishing of finely crafted wood boats.

U.S. Coast Guard (Boating Statistics, 1976) presents the following data in reporting fires and casualties on fiberglass reinforced polyester boats:

Total involved in accidents (fire or explosion of fuel)	294
Other fire or explosion	37
Fatalities	7
Equipment failures related to fires	121
Operator related	22

As noted, the number of boats in operation throughout the United States (at the end of 1975) approached 9 million. The fraction of fiberglass boats in accidents involving fire and including fuel explosions appears to be a somewhat insignificant 331/9,000,000 or 0.0036 percent. Further, the number of fatalities (7) resulting from fire-related accidents on fiberglass reinforced plastic boats is an even less significant number in comparison with the total number of boats and their operators in the year 1975.

Despite these relatively low fire-related fatality and damage rates, the replacement of the generally used non-fire retardant grade by presently available fire retardant resins can substantially reduce the flammability characteristics of such vessels with some increase in raw material cost because of the higher cost of the fire retardant resins. Flame spread of more than 500 for the general purpose resins (as measured by the ASTM E-84 test) is reduced to a range of 25—50 for the fire retardant resins. The improvement in fire safety obtainable in this way may be judged by the fact that red oak has a flame spread of 100 on this scale. The price premium necessary for the above improvement in fire safety can be approximated as follows.

The percentage increase in cost to obtain the benefit of a structure with a flame spread rate of 100 can be estimated as follows:
● A "typical" 30 ft. glass reinforced polyester boat contains 4,500 lbs. of general purpose polyester resin (exclusive of glass) having a density of 1.04 g/cm and selling for 35 cents a pound or $1,575 per boat.

- To achieve a flame spread rating of 100 would require the use of 5,538 lb. of a bromine-modified resin with a density of 1.28 g/cm and selling for 67 cents a pound or $3,710/boat. (Note increased weight of resin required because of greater density of the fire retardant material as well as the higher unit price).
- The selling price of a "typical" 30 ft. boat fabricated from general purpose resin is $40,000. The incremental cost ($2,136) due to improved fire safety is 5.34 percent of the current selling price. Typical trade mark-up formulas will cause this increase to be larger, perhaps 10.7 percent, resulting in a new price of $44,300.

It should be emphasized that the increase in sales price as estimated above will vary depending upon the size of the vessel. The price premium could be significantly larger for smaller boats where the raw material cost will be larger relative to the total manufacturing cost and the price/safety trade-off may not justify the use of the higher price raw material.

An analysis of the industry reveals that few builders employ "fire retardant" resins in the manufacture of their boats, Only one manufacturer displays the Underwriters Laboratory Classification Label (U.L.) indicating improved fire performance.

9.3 Operations

As many pleasure boats are lost to fire on shore, in storage, and in transit as are lost in the water. The following discussions involve the specialized precautions appropriate to each activity in which a boat is involved.

9.3.1 Construction and Repair

While under construction, a craft is probably more vulnerable to fire than at any time in its useful life. The flammability of polyester resins and the technology of their use in boat manufacture and repair require special precautions if fires are to be prevented. Since this committee has not included construction methods in its consideration, the reader is referred to NFPA No. 312, Fire Protection of Vessels During Construction, Repair and Layup, for detailed standards.

9.3.2 Fueling

Since so many boat fires occur before or during fueling, the following are some guidelines to be observed during fueling operations (NFPA No. 302, Fire Protection Standard for Motor Craft, Sect. 4-3).

Before Fueling
1. Stop all engines and auxiliaries.
2. Shut off all electricity, open flames and heat sources.
3. Check bilges for fuel vapors.
4. Extinguish all smoking materials.

During Fueling
1. Maintain nozzle contact with fill pipe.
2. Avoid overfilling.
3. Wipe up spills immediately and wash down.

After Fueling and Before Starting Engine
1. Inspect bilges for leakage or fuel odors.
2. Ventilate until odors are removed.

9.3.2.1 Small Craft Refueling Fire Scenario

A 19 ft. fiberglass motor boat is tied up at the fueling pier of a large marina on a warm summer afternoon. The owner-pilot shut down the ignition on the large outboard engine and he and two passenger guests stepped ashore, as he gestured for the attendant to fill his two 10-gallon gasoline tanks. That business was completed and the owner boarded and proceeded to start the engine.

The owner should have noted and accounted for two things as he hastily capped the gas cans, one of which overflowed (on his hands) and ran down inside the boat adjacent to the battery. First, the overflow should have been thoroughly flushed out with water. Second, attention should have been directed to the battery installation which in this case had a loose cable connection. Outboard boats are most often trailer oriented, towed to and from the owner's home, and the battery is frequently removed, to be installed at the time of the next boating activity. This accounts for some wear on the battery connections and can lead to carelessness in tightening the hold-down nuts.

The safety violations noted above led to a series of events which resulted in the total destruction of the boat within minutes. The gasoline fumes from the spill settled to the bottom of the boat adjacent to the battery where they were ignited by a spark generated across the loose battery terminal at the moment that the ignition switch was closed. The resulting explosion not only split the deck but ruptured one of the fuel storage cans releasing the gasoline to add fuel to the resulting fire. In a few seconds the flammable resins in the boat structure were ignited, thus intensifying the fire. The rapid fire spread caused the alarmed dock attendant to set the boat adrift where it burned to the waterline and sank less than 30 minutes after the initial ignition.

9.3.3 Storage

NFPA No. 303 provides guidance for fire safety of berthing and storage facilities at marinas and boatyards. These advisories for seasonal storage include:
1. Inspection for hazardous materials or conditions.
2. Removal of loose combustibles to suitable lockers.
3. Removal of galley fuel supplies.
4. Removal of lead-acid batteries or other precautions, if that is not possible.

9.3.4 Operational Fires

Fire ignition problems in boat operations are generally due to:
1. Fuel whose flammable vapors can be ignited by the hot engine or spark from the ignition.
2. Portable space heaters or cooking stoves igniting adjacent bulkheads or textiles contained in the accommodations area.
3. Cigarettes and smoking materials igniting fuel vapors, upholstered furnishings, mattresses and carpets.
4. Electrical overloads or sparks igniting fuel vapors, bulkheads and textiles.
5. Lightning strikes.

The areas of the boat most important to fire safety include:
1. Engine compartment, with its engine heat, electrical spark, and fuel vapor concerns.
2. Galley, with its electrical appliances and cooking flames.
3. Accommodation space, with its flammable textiles and cushions, cigarette smoking, and alcoholic beverages.
4. Structural hull, if made of fiberglass reinforced polyester or wood.

9.4 Polymeric Materials Used in Boats

9.4.1 Introduction

Polymeric materials are selected for use in boats because of their availability, cost and performance properties (e.g., function, mechanical durability, rot resistance, aesthetic appeal, ease of maintenance, low weight, and ease of fabrication). In the cost dominant marketplace in which these vessels are sold, little attention is paid to fire safety performance, in part because there is little information readily available to marine architects on the fire performance of polymeric materials.

The specific applications in which organic polymeric materials may be used on boats are:

Structural and Mechanical Furnishings/Carry-Ons

Structural and Mechanical	Furnishings	Carry-Ons
Boat hull and bulkheads	Furniture	Clothing
Bulkhead sheathing	Mattresses	Sporting Equipment
Overhead sheathing	Pillows	Aerosols
Decks	Drapes	TV-Radio
Electrical insulation	Curtains	Paper
Thermal insulation	Blankets	
Acoustic materials	Carpeting	
Fittings	Life Jackets	
Ducts		
Paint		
Adhesives		
Caulks		

The applications of materials for the above categories are detailed in Section 9.4.3.

9.4.2 Materials Evaluation Criteria

The various fire performance criteria cited throughout this volume (i.e. resistance to ignition, ease of extinguishment, heat flux, slow flame spread, etc.) should all be considered in selecting materials for boats.

The criteria are imprecise and depend on imperfect test procedures. Allowance has to be made for actual conditions of use: geometry, orientation, proximity of other materials and ignition source.

9.4.3 Material Applications

A review of materials in common use for the various applications has disclosed that:

1. Certain materials are adequate from a fire safety point of view.
2. Certain materials need improvement and
   (a) Adequate alternates are economically available, or
   (b) Alternates available are expensive, or
   (c) There is no alternate commercially available to do the job better.

9.4.3.1 Structural and Mechanical Components

9.4.3.1.1 Hull:

The most common basic structural materials in order of use are:
   (a) Fiberglass reinforced polyesters
   (b) Aluminum
   (c) Wood, such as, mahogany, oak, and plywood.
The statistics presented at the front of this chapter indicate few fire safety problems for aluminum hulls.

Wooden boats had been traditional until the fiberglass boat industry emerged, with U.S. Navy support, following World War II. Volume 1 of this series covers the state-of-the-art of flame retardant wood formulations; unfortunately, the impregnants are generally water leachable.

Fiberglass reinforced polyester is used as a basic structure for hulls; and for "glassing-in" composite structures such as fir lumber-foamed core composite reinforcing stringers, plywood foredecks, and balsa core composite overheads. As mentioned earlier, these unmodified glass reinforced resin composite structures have flame spread ratings of 500 or more, as measured by ASTM E-84 test (a value which indicates that they burn five times faster than red oak). By contrast, using reliable flame retarding procedures, fiberglass reinforced boats have been made routinely with flame spread ratings of less than 100.

It should be noted that formulations with flame spread ratings less than 25 (burning at one-fourth the rate of red oak) are attainable. At least one manufacturer is delivering boats which are flame retarded to attain flame spread ratings of 70—100. When one considers the relative cost of the improved formulation to the total investment in a furnished boat, it is reasonable to project that a favorable ratio of benefit to cost might be attained. This would justify the use of the less flammable but more expensive fire retardant resins at least in the larger vessels where the resulting increase in raw material cost is a small part of the total manufacturing cost.

9.4.3.1.2 Bulkhead and Overhead Sheathing and Decks

Bulkheads, ceilings, decks, counter tops, cabinet doors, etc., make use of plywood, oiled teak lumber and veneer, fabric backed wood veneer, alkyd paints, varnish, polyester gel coat, and high-pressure laminates.

For high-risk locations, such as cabinets near galley ranges, high pressure phenolic or melamine laminates offer superior flammability characteristics. So do aluminum panels, vinyl clad aluminum panels, and fire retardant polyester panels. The substitution of these low-flammability materials for wood in these high hazard locations could do much to improve fire safety and reduce fire hazards on these boats.

Deck coverings of vinyl asbestos tile would meet almost any fire hazard requirement on boats.

While the committee is not aware of the use of thermoplastic structural foam panels on boats, e.g., in simulated carved wood applications, the rapid introduction of such panels in buildings in recent years and the consequent severe rise in fire hazard dictates to the appropriate boat regulating authorities a need for careful monitoring for such fire hazards in the future. The committee would recommend a temporary ban on the use of such materials on boats until the potential fire hazard can be assessed.

9.4.3.1.3 Electrical Insulation

Electrical cables do not present the same hazard on boats that they do on larger and military vessels. Polyvinyl chloride wire jacketing would be acceptable for the most stringent needs anticipated.

9.4.3.1.4 Thermal and Acoustic Barrier Materials

Fiberglass or glass wool is generally acceptable. Urea-formaldehyde foam is also cost competitive, functional, reasonably fire safe, and has the advantage of being injectable into existing cavities. It is, however, friable, and cannot be used in exposed locations.

Flexible urethane foams are unacceptable by all six flammability criteria listed in section 9.4.2. Rigid urethane foam formulations may find acceptance if they are specifically certified by appropriate regulatory agencies using recognized test procedures.

9.4.3.1.5 Fittings

Small molded parts from most commodity plastics represent little hazard in this application.

9.4.3.1.6 Ducting

Aluminum and coated steel sheet metal ducting for air handling and ventilation are acceptable materials. Flame retarded fiberglass resins and phosphate plasticized PVC are also suitable. Rigid urethane foam ducts could possibly be flame retarded to be acceptable, but this has not yet been demonstrated.

9.4.3.1.7 Adhesives and Caulks

These materials are not generally used in sufficient quantities to be of concern. Contact adhesives meeting Federal Specification MMM-A-130 are acceptable where requirements are stringent.

9.4.3.2 Furnishings

There is no control of furnishings brought on board by boat owners.

On boats with wood or polyester-fiberglass hulls, the presence of limited quantities of additional materials with less than optimum fire resistance properties might have a negligible effect on the total hazard, unless these additional materials are highly flammable, like structural foam furniture, or are located in critical areas where they might serve as kindling for the destruction of the entire boat.

Certain materials present the greatest potential for ignition or the speed of fire because of their composition, location, or sheer bulk. The following are considered high-risk materials:

9.4.3.2.1 Combustible Textiles

9.4.3.2.2 Furniture, Mattresses, and Pillows

Furniture upholstered with rayon or cotton fabrics, alone or in blends with any other fiber or blend of fibers are to be avoided because all untreated cellulosic upholstery fabrics ignite easily and spread flames rapidly. Alternates, flammable but with lower fire risk than cellulose, are fabrics of polyester, nylon or wool, or certain PVC cloth laminates. To achieve a high degree of fire safety, at higher cost, Beta fiber glass cloth, Nomex aromatic polyamide, and Kynol® Nomex® blends are recommended aboard U.S. Navy vessels where authorized.

Flexible urethane foam is used almost universally for cushioning in small private boats. Urethane foam cushions are a major contributor of fuel to fires in every type of occupancy. Urethane foam continues to be used because of its superior mechanical properties, price, and ease of fabrication, but a fire-safe competitive alternative is sorely needed. Polychloroprene foam rubber (Neoprene) is a relatively expensive fire-safe alternative used by the U.S. Navy. Recently DuPont has offered a thin sheet foam material containing polychloroprene and alumina hydrate to serve as a fire barrier encasement for urethane foam cushions. Boric acid-treated cotton batting is a less aesthetic alternate which loses its effectiveness in a wet environment.

9.4.3.2.3 Drapes, Curtains and Blankets

Textile fibers and their flammability are covered in Chpater 4 of this volume. Cotton and rayon and their blends with thermoplastic fibers (nylon, polyester, acrylic) are to be avoided because of ease of ignition and high-rate burning. *Flame-retarded* cotton and rayon formulations are considered more acceptable. Polyester and nylon are flammable but burn slowly; however, the small flame front is intensely hot and could ignite other adjacent surfaces readily. Modacrylics do not ignite readily but have functional shortcomings. The Navy specifies fibrous glass, aramid (Nomex®) or Kynol® Nomex® blends for most applications when a specification is used.

9.4.3.2.4 Carpeting

This topic was discussed in detail in Chapter 4. The federal standard, DOC-FF-1-70 (pill test), eliminates from sale all carpets that might ignite from a cigarette or other similar small ignition source. Among the commodity carpet fibers, the following ranking was deduced in Volume 7:

Most Difficult to Ignite:	Wool or polypropylene
	Nylon
	Polyester
Easiest to Ignite:	Acrylic

Slowest Burning Rate (Radiated at 6.2 watts/cm^2):

Polyester, nylon or polypropylene	$\sim$ 2.5 cm/min.
Wool or acrylic	$\sim$ 7.5–10.0 cm/min.
Viscose	$\sim$ 125 cm/min.

Other work described in Volume 7 showed that carpet laid directly on the floor burns less readily than carpet on underlay. All carpet evaluations are based on horizontal use on the floor; vertical mounting on a wall is an extremely dangerous fire hazard; vertical mounting should be banned. When it specifies, the U.S. Navy specifies Beta glass or Nomex for carpet; alternate floor coverings are vinyl asbestos tile and conforming vinyl sheet.

9.4.3.2.5 Life Jackets

Neoprene coated nylon or urethane coated nylon jackets are generally available. There are no fire safe alternatives for critical applications. Possible future developments are flame retardant neoprene coating on nylon or aramid fabric.

9.4.3.3 Carry-Ons

The materials carried on board by passengers are generally easier to ignite and more flammable than the structure and permanent contents of the boat. There appears to be no way to reduce this fire hazard except by better education of the boating public, by improved discipline, and with better storage facilities. Relative to a similar problem in the more stringent commercial aircraft industry, this problem is not severe. A partial solution would be to provide proper stowage facilities.

9.4.3.4 Interior and Furnishings

For the boats considered in this chapter, there are no regulations with respect to interior and furnishings beyond those that apply to all consumer products.

9.4.3.4.1 Textiles

The earliest textile flammability standard appears in the text of the flammable Fabrics Act of 1953. This test was designed to eliminate from the market place those fabrics which were most susceptible to flash fire (e.g., brushed rayon sweaters). It is rare to find a textile today that does not pass this test.

9.4.3.4.2 Carpets

The first standard to be issued pursuant to the 1967 Amendment to the Flammable Fabrics Act was DOC-FF-70-1, the "pill test" for carpets. This test, designed to simulate the hazard of a dropped cigarette or fireplace ember, caused significant changes in the structure and composition of carpets.

If a carpet passes this test, it does offer greater resistance to the conditions it simulates, but it offers no guarantee of performance in a cabin-size fire. A more stringent and realistically designed carpet flooring radiant panel test has been developed by NBS. To date, it has not been adopted for any regulatory standard. More details on carpet flammability appear in Volum 7, Buildings, of this series.

9.4.3.4.3 Mattresses

A federal standard DOC-FF-4-72 is in effect for all mattresses sold in U.S. The standard improves the resistance of a mattress to glowing cigarette ignition.

Initially Standards under the Flammable Fabric Act (FF) were developed by the National Bureau of Standards and were released by the Department of Commerce.

9.4.3.4.4 Upholstered Furniture

A federal standard similar to the mattress standard is under development and is close to adoption.

9.4.3.4.5 Interior Finishes

While no standards exist, the whole spectrum of tests available for the fire hazard classification of building materials is available to be called upon when the need is identified.

9.5 Test Methods and Standards

9.5.1 Structural and Mechanical Components

The U.S. Coast Guard has the responsibility for development of fire protection standards to minimize the incidence and consequences of commercial and private shipboard fires, consistent with economic considerations.

It is common knowledge that pleasure craft with fiberglass reinforced plastic hulls, once ignited, burn very rapidly and completely. Nevertheless Boating Statistics (1976) shows minimal casualties as a result of fires (including fuel fires), and annual property damage of about $3.3 million for a fleet of 9 million boats. In the face of such statistics, there appears to be no basis for imposition of fire protection standards which would cause more than nominal expenditures to achieve compliance. Thus, the absence of any Coast Guard structural materials standards for pleasure boats appears justifiable by the accumulated statistics.

The state of the art with respect to fiberglass reinforced polyester is such that significant fire safety improvements are now attainable at relatively small incremental cost (about 10 percent of sales price for the larger boats). Therefore, the accident statistics should be followed on an annual basis; any significant increase in casualty or property loss should be taken as a signal of the need for regulatory action.

Small passenger vessels carrying between 6 and 150 passengers, not in international voyages, and which fall within the scope of this chapter, are subject to Federal regulation. These vessels traditionally were constructed of wood; however, recently construction has switched to fiberglass reinforced plastics (FRP). In 1972 it was recognized that FRP hulls constructed of general purpose resin presented a potentially serious problem in a vessel carrying this number of passengers. Because wood had been the mainstay in the construction of this type of vessel, the fire properties of wood were utilized as a benchmark for the requirements for FRP hulls. By requiring hulls to be constructed of fire retardant resin, complying with MIL-R-21607, the fire hazard properties (i.e., ease of ignition, spread of flame, and heat content) closely approximate those of wood. Coast Guard Regulation, Title 46, Code of Federal Regulations, CFR 177.10-5 requires this type of vessel to utilize MIL-R-21607 resin if FRP is utilized as a primary structural element.

Despite this attempt to limit the fire hazard to the equivalent of wood, the state of the art of the unsaturated polyesters technology can realistically require the attainment of flame spread ratings of 25 or less by (ASTM E-84) with only moderate (10—15 percent) increase in overall cost. It would seem desirable to reduce the allowable flame spread ratings from 100 to 0—25 to take advantage of the improved state-of-the-art. Although such a change in regulations could be expected to reduce the fire hazard in these vehicles, it would be desirable to confirm this supposition by a series of full scale and mock-up fire tests.

9.6 Conclusions and Recommendations

Conclusion: The casualty and property loss data accumulated by the U.S. Coast Guard for registered boats show a very low injury rate and small property loss from fires/explosions. Most of the reported fires occur in boats with wood or fiberglass-reinforced hulls and inboard engines. *Recommendation:* The low incidence and consequences of fires aboard boats do not at this time appear to support the imposition of fire protection regulations that would add to the cost of recreational boats.

Conclusion: The state-of-the-art of fiberglass-reinforced plastic technology is such that products with greatly improved fire safety performance are available at a moderate price premium. The fire safety of interior surfaces and furnishings can be improved substantially by state-of-the-art technology. *Recommendations:* Promote the availability of materials with improved fire safety performance. Carefully monitor the Coast Guard casualty data in order to identify potential fire hazards requiring new safety standards. Substitute less flammable high-pressure phenolic laminates or coated aluminum panels for wood and wood veneer in high-hazard bulkhead and overhead paneling.

REFERENCE

Boating Statistics 1975, CG-357, Department of Transportation, U.S. Coast Guard, Washington, D.C. 1 May 1976.

CHAPTER 10

GENERAL CONSIDERATIONS

10.1 Introduction

Fire has been the constant companion of water-borne transportation over the ages. Ships were constructed of and outfitted with highly flammable materials; the prevalence of open fires for cooking or repair made fire a commonplace occurrence. The advent of all-steel hull construction during the 19th century for the first time made the construction of a truly fire resistant ship possible. However, the continued extensive use of flammable materials for interior furnishings and decorative finishes has so far postponed the advent of a truly fire safe vessel.

Fire safety in water-borne vessels has been approached in a more regulated and systematic way than for land transport and buildings. This may, perhaps, be because of a series of disastrous passenger ship fires in the 1930s. In 1934, as a result of the Morro Castle fire in which 124 lives were lost, Congress passed the Materials and Methods of Construction Act, which charged the U.S. Coast Guard with the regulation of all U.S. Merchant Marine vessels. A corollary to these events was the definition of a unified approach to maritime fire safety. This philosophical approach to marine fire safety not only severely limited the use of combustible materials in ship construction, but required that the vessel be divided into internal compartments constructed entirely of fire resistant materials such as steel so as to contain any shipboard fire to its origin. This philosophy of fire control was advanced by the United States and subsequently made one of the three acceptable methods of fire control in the 1948 International Convention for the Safety of Life at Sea (SOLAS, 1948). Although SOLAS '48 recognizes two other fire safe ship construction methods (see Appendix A Chapter 7), U.S. regulations permit only the above method for use in the construction of United States flag vessels.

Naval ship construction has been generally all-metal for more than a century, during which time diligent efforts have been made, consistent with existing technology to reduce or eliminate the incidence of shipboard fires. Despite this diligence, the elimination of shipboard fires has been an elusive goal.

In contrast to the closely controlled and regulated nature of naval and merchant marine ship construction with its emphasis upon fire safety, the rapidly multiplying private pleasure craft remain essentially unregulated for fire safety. Because of this lack of regulation, all kinds of polymers are being widely used in small boat construction, with little knowledge or understanding of their flammability or their effect on fire safety. Reinforced polyester lamintes dominate the market because of their low cost, ease of fabrication, ease of maintenance, and resistance to rot and corrosion. The flammability of these materials may very well present the single greatest future fire safety problem in this area. This problem is discussed in more detail in Chapter 9.

Despite the great societal benefits pressuring the expanded use of polymers in ship construction, the ready flammability of most of the low-cost high-volume materials produced at the present time makes the close control of these materials desirable; but, as noted earlier, the low accident rate does not warrant immediate drastic regulation (9.5.1 and 9.6.2).

10.2 Impact of Polymeric Materials on Ship Construction

There has been up to now the relatively small impact of polymers upon ship construction because of the tight safety regulations already described. However, the economic pressure remains great for the use of these newer and inherently advantageous materials in vessels. Some of the more important of these advantages are reduced initial cost, reduced weight, and reduced maintenance costs (derived from their superior resistance to corrosion and soiling). The use of these materials also leads to significantly reduced fabrication costs and reduced energy costs both in materials production and systems operations.

10.3 Energy Considerations

The energy requirements for the production of a unit of polymer are only one tenth to one half that for such metals as aluminum or magnesium while energy savings of 10–80 percent are possible relative to steel, depending upon the polymer chosen (Brake, 1976). This energy advantage of polymers over metals also extends into energy consumption, particularly in high-speed, special purpose vessels, because the lower density of polymers reduces the weight of the finished product with a resulting saving in energy needed to propel the ship through the water.

10.4 Fire Protection Philosophy

As already noted, U.S. flag merchant vessels must conform to a fire safety philosophy that was an outgrowth of the Morro Castle fire. Basically, the construction philosophy embodies the following principles (Sheehan, 1976):
   a. Protection of the means of escape.
   b. Limitation of the amount of combustible material used in construction.
   c. Containment of the fire within the space of origin by appropriate fire resistant construction.
   d. Isolation of passenger and crew accommodations from fires in the cargo and machinery spaces.

The success of this method of construction has been amply demonstrated by the absence of major fires on U.S. flag merchant vessels since the generation of the regulations. Unfortunately, the SOLAS 60 convention allows the construction in countries other than the United States of vessels conforming to other methods that utilize alarm and sprinkler systems to protect a less expensive but more combustible structure (International Convention, 1976). By treaty, these more combustible vessels are allowed to operate freely in the United States ports and waters. They carry both American passengers and cargo. In this respect, U.S. citizens and property on board are exposed to greater hazards than exist on U.S. vessels. Since most cargo and passengers in international commerce are carried in vessels of foreign registry, American citizens are essentially denied the protection of an apparently safer method of ship construction. In this regard, it should be noted that there are no large passenger ships of U.S. registry in operation today. The added cost of providing fire safety has been a contributing factor to the disappearance of American passenger ships.

10.5 On-Going Fire Safety Research

Both the Coast Guard (Sheehan, 1976) and the Department of the Navy (Executive Summary, 1976) maintain significant research programs aimed at defining the important fire safety variables and methods of improving shipboard fire safety. Perhaps the best example of such fire safety research is the large-scale test program being carried out at the Coast Guard Shipboard Fire and Safety Test Facility — Mobile, Ala. Here, compartment fire tests are carried out on the 8,500-ton T-1 tanker MV Rhode Island. This facility is available for both industrial and government use. Examples of the type of research recently carried out at this facility are fire extinguishing systems research, research on modular fire protection systems, research for container ships cargo fire protection, and fire research tests to determine the equivalency of materials of construction. The Navy Department also has a vigorous program on shipboard fire and safety research because of its importance to overall fleet efficiency both in peace and in war. This program (Executive Summary, 1976), which is largely carried out at the Naval Research Laboratories, consists of extensive research in a wide variety of ship-related subjects including flammability of materials, smoke generation during combustion, and the toxicity of gaseous combustion products of shipboard materials including synthetic polymers. In addition, naval safety research includes studies in improved fire fighting systems, novel means of smoke abatement, detectors and alarms, fire retardant intumescent paints and mastics, and fire barriers, to list only a few specific examples. Such fire research programs are necessary to maintain and improve the current high degree of fire safety currently built into U.S. flag oceangoing vessels.

Navy Department fire research is concentrated in three major areas previously identified as requiring improvement (Executive Summary, 1976):

1) Fire Dynamics and Scaling
2) Fire Effects on People, Including Toxicity Effects
3) Materials Characterization and Improvement in a Fire Environment.

These programs include research on lifesaving equipment and devices, fire protective clothing, chemical extinguishing agents, fire detectors, fire resistant and low-smoke materials, and firefighting methods. Continuing programs such as these have had a significant effect in reducing and controlling fires on naval combat vessels and shipping.

10.6 Educational Aspects of Polymer Flammability

The fire safety problem in the United States, as elsewhere, has not been approached on a systems basis. Some desultory efforts and expenditures have been made on a component basis or in response to a major fire calamity. Most efforts have been directed toward "fire drills" for escape, spot extinguishment of flames (sprinklers, CO_2 flooding systems), and application of "improved" materials to decrease the danger of ignition. The support of governmental and private groups has been fragmented and applied without useful priorities based on risk analysis.

Only recently has there been national awareness of the increasing severity of the fire problem and an approach made to an attack on a systems basis (National Fire Center, RANN support, etc.); but funds and progress have been limited.

In the past there has been little public knowledge of or training in basic fire safety technology. Until recently there has been no place to receive such training. Today there are only a few colleges, technical schools, or other institutions where formal training in one or more aspects of fire safety can be obtained.

In general, there has been no basic methodological way to approach fire safety. Partial approaches, sometimes self-defeating, are found in construction codes, building specifications, governmental regulations and guidelines, and insurance requirements. These, more often than not, were developed from analysis of a catastrophe or in response to a specific need.

This committee has already commented in some detail on these matters (Volumes 4 and 6 of this series) and indicated a national fire safety need for:

> Education and training.
> Direction and coordination of efforts.
> Financial support.
> Communication.

Another significant contribution to the fire safety problem is the general lack of knowledge and appreciation of polymer fire safety among the general public and even among many practicing engineers, architects, and designers. Many synthetic polymer materials burn differently from the more familiar natural ones. They may melt and drip and often give off dense and acrid smoke. Although some are less flammable than such familiar materials as wood or cotton, others burn with more intense flame and resist the efforts of conventional firefighting. Such behavior can lead to damage and loss of life which might have been avoided with a better understanding of the performance of materials in a fire. Better understanding of such fire hazards can be promoted through educational efforts (National Commission on Fire Prevention and Control, 1973; Tabor, 1975).

This general lack of understanding of polymer properties may result from the fact that polymer science and engineering are only 30—40 years old, and school training in these relatively new materials has been generally slow at both the secondary and collegiate levels. Salamone, Deanin, Young, and Pearce (1973), among others, have briefly discussed the reasons, as well as having tabulated information on polymer science and engineering education in the United States.

The general lack of understanding of polymer flammability among the public has had a serious impact upon shipboard fire safety as in some other polymer application areas. But a wider understanding of the subject among naval and marine architects and engineers would undoubtedly assist in improving shipboard fire safety, at least indirectly, by leading to improved design and application of polymers in ship and boat construction. Incorporating general courses on polymeric materials in science and engineering educational curricula would appear to be in order to correct this educational deficiency.

10.7 Conclusions and Recommendations

Conclusion: The fire safety of United States Flag seagoing shipping, both naval and merchant, is well controlled in general. *Recommendation:* Improved polymer fire safety education would indirectly lead to improved shipboard fire safety. *Recommendation:* The fire safety aspects of polymeric materials should be included in the curricula of naval architects and marine engineers.

REFERENCES

Brake, P. F., "Potential Tank-Automotive Applications for Organic Materials, Part II Composite Materials" Army Materials and Mechanics Research Center Report, March 31, 1976.

Executive Summary, First Navy-Wide Workshop on Shipboard Fire Protection Research and Development, Naval Research Laboratory, March 9–10, 1976.

National Commission on Fire Prevention and Control, "America Burning" U.S. Government Printing Office, Washington, D.C.: 1973.

1960 International Convention for Safety of Life at Sea.

Salamone, J.D., Deanin, R.D., Young, M.G., and Pearce, E.G.J., Chem. Educ., 50, 769 (1973).

Sheehan, D. F., Presentation to the NAS Committee on the Fire Safety Aspects of Polymeric Materials, November 30, (1976).

Tabor, J.K., Commerce Today, 5 (15), April 28, 1975.